THE BLACK SEA IN CRISIS

RELIGION, SCIENCE AND THE ENVIRONMENT

Symposium II:

THE BLACK SEA IN CRISIS

An Encounter of Beliefs: A Single Objective

20 – 28 September 1997

Under the auspices of
His All Holiness The Ecumenical Patriarch Bartholomew I
His Excellency Mr Jacques Santer, President of the European Commission

Editors

Sarah Hobson
Open Channels, Oundle, UK

Laurence David Mee
Plymouth Environmental Research Centre, University of Plymouth, UK

Assistant Editor

Sally Morgan

World Scientific
Singapore • New Jersey • London • Hong Kong

Published by

World Scientific Publishing Co. Pte. Ltd.
P O Box 128, Farrer Road, Singapore 912805
USA office: Suite 1B, 1060 Main Street, River Edge, NJ 07661
UK office: 57 Shelton Street, Covent Garden, London WC2H 9HE

British Library Cataloguing-in-Publication Data
A catalogue record for this book is available from the British Library.

THE BLACK SEA IN CRISIS

ISBN 981-02-3769-3

Printed in Great Britain by J. W. Arrowsmith Limited, Bristol

Courtesy of Nicolas Manginas

His All Holiness the Ecumenical Patriarch Bartholomew I and His Beatitude Patriarch Catholicos Elias II of All Georgia locating some of the problems facing the Black Sea.

Courtesy of Nicolas Manginas

Symposium participants in plenary.

Courtesy of Nicolas Manginas

Discussions on the environment between the Ecumenical Patriarch Bartholomew I (second from right) and Patriarch Alexy II of Moscow and All Russia (second from left) with other members of their churches.

Courtesy of Jill Purce

Professor David Bellamy (right) with Bishop Mauro Morelli of Brazil.

Courtesy of Nicolas Manginas

A traditional welcome with bread and salt for the Ecumenical Patriarch Bartholomew I at Odessa.

The threatened bounty of the Black Sea.

Sailing in the Black Sea.
Courtesy of A. Poulidis, Oxygen

Courtesy of Jill Purce

Waiting for the Ecumenical Patriarch.

Foreword by the Religious and Scientific Committee

'If I am the only person who dreams it is only a dream, but if several of us live the same dream it can become a reality.' - Brazilian proverb

This book records the second of a series of symposia on the general theme of 'Religion, Science and the Environment'. An objective of the first symposium was to bring the insights, knowledge, inspirational abilities, and methods of religion and science together to ensure that the natural world on which all life depends is protected from progressive deterioration, while human welfare increases. The second symposium became the natural place to identify specific actions that could be taken, with a particular focus on the Black Sea as a region and environment in crisis.

Held under the auspices of His All Holiness the Ecumenical Patriarch Bartholomew I and His Excellency Mr. Jacques Santer, President of the European Commission, the second symposium took place from 20 to 28 September 1997 on board the F/B *E. Venizelos.* Over 400 participants were transported from the Black Sea ports of Trabzon, Batumi, Novorossiysk, Yalta, Odessa, Constanta, the Danube Delta and Varna, and on to Istanbul. The journey of over 1,000 nautical miles, through the lands and seas of 2,500 years of Greek colonies and the Roman, Byzantine and Ottoman Empires, ended in Thessaloniki, the cultural capital of Europe for 1997.

In his introduction to the symposium, Richard Chartres, Lord Bishop of London, reminded participants that:

> after the great flood described in the story of Noah, the threat of divine destruction of the earth was withdrawn and the rainbow was appointed as a sign of a 'covenant between God and every living creature of all flesh that is upon the earth.' Now it is a human rather than the divine threat to make the planet 'waste and void', which causes anxiety. This is the impulse which is bringing together scientists, people of faith and others concerned for the future of our planet to the fertile but heavily polluted Black Sea region.

The symposium brought together people from many different backgrounds and beliefs, including Christian, Jewish and Muslim religious leaders from the Black Sea region as well as representatives of faiths from many other parts of the world. All the Patriarchs from the Black Sea region were present. Leaders from the region included presidents, vice-presidents and ministers of the environment, while prominent scientists participated bringing expertise at local, regional and international levels.

The symposium presented an opportunity for the Ecumenical Patriarch to visit his Church in five countries and to discuss his concerns about the Black Sea environment. He was able to emphasise the centrality of this issue and the leadership role of the Church. The symposium was also able to raise awareness among important religious and political leaders of the region and the European Union about the environmental crisis of the Black Sea and the threat from human activities.

One practical result of the symposium was the identification of the need to maintain the Black Sea Environmental Programme as a critical institution for implementing a plan of action for the region. At the same time, a broader understanding of the nature of the problem was identified and emphasised, specifically the contribution of the Danube River and its watershed to the environmental problems of the Black Sea. Intensive workshops held during the symposium also identified key actions for awareness and outreach, education, as well as capacity building and maintenance, including rebuilding the scientific capacity of the region. Interest in supporting capacity building was expressed by representatives of the European Union, the World Bank, and the Global Environmental Facility who were aboard the ship. Influential individuals who participated in the symposium also offered their help.

Specific educational actions for the region were identified. As a consequence of the symposium, plans have been developed for expanding and deepening the understanding of clergy on environmental problems and solutions. Teachers need opportunities and training for the informal exchange of information on environmental issues, while more formal meetings among ministers of the environment and education would support the development of educational materials and environmental education on the Black Sea.

Beyond specific proposals and with long-term significance has been the creation of a network of concerned individuals from the worlds of religion, science and the environment, whose collective creativity, energy, and influence can be used to amplify many times the inspiring message and mandate provided by the Ecumenical Patriarch. The use of this network of people to accomplish the goals of the symposium is our opportunity. Determining how to do this effectively, guided by scientifically and ethically sound principles, is our challenge.

Major financial support for the symposium was provided by the Greek Ministry of Culture, the city of Thessaloniki, the Hellenic Telecommunications Organisation, Titan Cement Company, and an anonymous donor. The Religious and Scientific Committee that organised the content of the symposium was once again inspired by a remarkable woman, Maria Becket, without whose vision, energy, and powers of persuasion, the symposium would have never occurred.

The Most Revd. Metropolitan John of Pergamon
Professor Daniel Amit
Dr. Zaki Badawi
The Rt. Revd. and Rt. Hon. Richard J.C. Chartres
Dr. Charles N. Ehler
Dr. Graeme Kelleher
Professor Jane Lubchenco
Professor Laurence Mee
Professor Erdal Özhan
Rabbi Naftali Rothenberg
Professor Yuvenaly Zaitsev

Editors' Note

In editing the material emanating from the symposium we had a number of issues to consider. First was the choice of material. The academic papers which were formally presented during the symposium provided the most substantial content and these form the first eight chapters of the book. Of considerable importance were the many proposals made in the course of the symposium during the workshops and elsewhere; these proposals were carefully analysed by the Religious and Scientific Committee after the symposium and the conclusions of this work are presented in a final section, Recommendations for Helping to Protect The Black Sea.

Then there was the matter of length, with individual papers ranging from 1,000 words to almost 5,000 words; there was also concern about the difficulty of including material that had been intended exclusively for a listening audience compared with material that had been prepared as text for reading. Finally, there was the issue of language, since a range of English styles was used, including American English, English English, Indian English and second-language English, not to speak of the difference arising from the language of various disciplines, for example, scientific, religious, philosophical and activist.

Each paper was therefore edited on the basis of a finished length of not less than 1,500 words and not more than 2500 words; retaining the main arguments of the material; honouring the style and integrity of the piece; correcting ungrammatical English but not altering the type of English used; and standardising the different forms of spelling. The edited papers were submitted to their authors for comment and approval.

We would like to thank the following people for their assistance: Sally Morgan for her work as assistant editor; Linda Sharp for help in editing the material; Lyn Parker for proof reading, Anne Langley for word processing and David Chadwick of Chadwick Associates for typesetting and design.

Sarah Hobson and Laurence Mee

July 1998

The publication of this book has been made possible
by the generous support of the

Michael Marks
Charitable Trust

Contents

Introduction

The Black Sea Today[1]

Laurence David Mee

This introductory paper examines the state of the Black Sea today and the efforts to protect it. It recognises that the environmental degradation in the Black Sea is a consequence of a larger problem, that of how humans perceive and value nature. This problem is certainly not limited to the countries around the Black Sea, though the many years of totalitarian regimes in most Black Sea countries have left a particularly challenging social and environmental legacy.

The crisis in the Black Sea is part of a crisis of values which may be planetary. Indeed our use of terms such as 'environmental crisis' provides a shield against the reality that the problems are of human origin and the solution is in our hands. Deepening our understanding of this crisis of values offers hope for sustaining our presence on this planet.

The situation in the Black Sea, Europe's most isolated marine area and its most degraded, is not hopeless. The current economic and social changes in Central and Eastern European countries open opportunities to introduce sustainable practices in order to overcome major problems such as eutrophication, overfishing, landscape and habitat destruction or pollution with solid waste and sewage. However, it is apparent that the driving force for such change must come from an understanding and empowered public as well as better informed and able state and private sectors. How can the change be made?

A casebook of symptoms of environmental degradation

It does not require a profound knowledge of oceanography to appreciate the unique nature of the Black Sea. Any physical map of Europe is sufficient to reveal its basic characteristics: its comparable surface area to the Baltic or North Seas; its huge drainage basin covering over one third of Europe and including major portions of seventeen countries; its great depth, over two kilometres in places; and its isolation. The map also shows the connection to the Mediterranean through the narrow, twisting Bosphorus, 700 metres wide and less than 60 metres deep in places, and through the Sea of Marmara and the Dardanelles. It does not reveal however, that the Bosphorus annually carries 600 cubic kilometers of surface water flowing from the Black Sea and 300 cubic kilometers of deep water replacing it from the Mediterranean. It also carries some 50,000 cargo ships (including 1,500 tankers) annually,[2] and flows through the middle of megalopolis Istanbul.

The Black Sea is also one of Europe's newest seas. It was formed a mere seven or eight thousand years ago when sea level rise caused Mediterranean water to break through the Bosphorus valley refilling a vast freshwater lake tens of metres below the prevailing sea level. The salty water sank to the bottom of the lake, filling it from below and forming a strong density gradient (known as a *pycnocline*) between the Mediterranean water on the bottom and the freshwater mixed with some seawater near the surface. The depth of this natural density barrier depended (and still depends) upon the supply of fresh water from rivers and rain, and the energy available from the wind and the sun for mixing it with the underlying seawater. The oxygen in the incoming water was quickly exhausted by the demands of bacteria associated with decaying biota and terrestrial organic material falling through the density gradient into the bottom water. Within a few hundred years, the Sea, below some 100 - 200 metres depth, became depleted of oxygen. The bacterial population switched to organisms capable of obtaining their oxygen by reducing dissolved sulphate to toxic hydrogen sulphide and the resulting water body became the largest volume of anoxic water on our planet.

For several thousand years therefore, only the surface waters, down to the 'liquid bottom' pycnocline, have been capable of supporting higher life forms. Though not very biologically diverse compared with open seas at similar latitudes, the Black Sea developed remarkable and unique ecosystems, particularly in its expansive north-western shelf where the sea is relatively shallow. The seabed in this part of the Black Sea was well oxygenated since it is well above the pycnocline. This area, and the adjacent shallow Sea of Azov, also receives the inflow of Europe's second, third and fourth rivers, the Danube, Dnieper and Don. A particularly unique ecosystem developed based on the 'keystone' benthic (bottom living) red algae, *Phyllophora* sp., which formed a vast bed with a total area equivalent to that of Belgium and The Netherlands. The term 'keystone' is not used lightly: like the keystone in the middle of a stone bridge, its removal causes the entire structure to collapse in a precipitous manner. This particular keystone was also a place of great beauty, vast underwater fields of red algae, home to a myriad of dependent animals, linked together in a complex web of life.

Despite its uniquely fragile natural physical and chemical characteristics, the Black Sea ecosystem appears to have been relatively stable. During the first half of the twentieth century, perhaps until three decades ago, there was little evidence of human impact on the Sea or on its flora and fauna. Some changes had occurred however, and these were precursors of much worse events to come. Sensitive monk seal populations, for example, began to decline from the late nineteenth century, driven from their breeding grounds by human activities. Nowadays the rarely sighted miniscule population of these seals seems certainly doomed. Indeed, there is no certainty that any of these animals remain in the Black Sea. Another early change was through the introduction of a number of exotic animal species, introduced by accident from the hulls, bilges or ballast tanks of ships, and which flourished to the detriment of the Black Sea's characteristic fauna. The voracious predatory sea snail *Rapana thomasiana*, for example, arrived from waters around Japan in the mid-1940s and devastated beds of the Black Sea genotype of the common oyster, *Ostrea edulus*. It is one of a list of some 26 species introduced through human activity (accidentally or intentionally) since the beginning of the century and which have profoundly altered the Black Sea ecosystem.[3]

Another gradual change was taking place on the coastlands of the Black Sea. Urban construction occurred in an unplanned and haphazard manner. The Black Sea was an increasingly popular tourist venue, particularly for the peoples of the former Soviet Union and the other Eastern and Central European COMECON countries. This, together with competing demands for space from shipping, industry and coastal settlements (mostly with inadequate waste disposal), placed increasing demands on coastal landscapes. The damming of many rivers brought hydrological changes, particularly through the decrease in sediment flux to the coast, a phenomenon that led to major problems of erosion.[4] This, in turn, was often ineffectively combated by the construction of a very large number of structures to protect beaches (groynes). These further degraded the landscape and exacerbated pollution problems. In the competition for coastal space, the natural environment was the seemingly inevitable loser. The human population has continuously encroached on the ecosystem that it is part of and upon which it depends.

From the late 1960s to the early 1990s, events occurred in the Black Sea that can objectively be considered as an environmental catastrophe.[5] The strongest single symptom of the catastrophe was the virtual elimination of the *Phyllophora* ecosystem from the Black Sea's north-western shelf in a matter of some ten years. The chain of events leading to the decline of this ecosystem started with an increase in nutrient flux down the major rivers, particularly in the late 1960s when fertiliser use increased markedly as a result of the 'Green Revolution'. This brought about a decrease in light penetration in the sea due to the increased intensity of phytoplankton blooms (eutrophication). Deprived of light, the red algae and other photosynthetic bottom dwelling (*benthic*) species quickly died. Their function was lost as a source of oxygen to the bottom waters of the shelf seas and as a habitat for a wide variety of organisms. The bottom waters of the north-western shelf became seasonally hypoxic (very low oxygen) and even anoxic (no measurable oxygen). Thousands of tons of benthic plants and animals were washed up on the shores of Romania and Ukraine and the seabed became a barren area with a very low biological diversity.

The loss of the north-western shelf ecosystem had an impact on the entire Black Sea. It also coincided with a period of expansion in the fisheries industry and the application of high technology fish-finding hydroacoustics and more efficient, though unregulated and destructive, purse seining and bottom trawling gear. The consequence was a decrease in the diversity of commercially exploitable fish species from some 26 to six, in less than two decades. As eutrophication advanced in the Black Sea, the smaller fish species such as anchovies and sprat were favoured since they depend upon the phytoplankton-driven pelagic ecosystem, rather than the benthic one. Furthermore, their predators had often been removed by overfishing or habitat loss. As a consequence, fishing effort switched to these lower value species. Annual catches of anchovy for example, rose from 225,000 tons in 1975 to some 450,000 tons a decade later.[6]

In the mid-1980s, another exotic species arrived in ships' ballast waters, the ctenophore *Mnemiopsis leidyi*, sometimes known as the comb jelly.[7] This species was brought from the eastern seaboard of America and, without predators, flourished in the eutrophic Black Sea environment where it consumes zooplankton including fish larvae. Perhaps the word 'flourished' is an understatement. At its peak in 1989-90, it is claimed to have reached a total biomass of about one billion tons (1,000,000,000 tons wet weight) in the Black Sea, more than the world annual fish harvest! This massive population explosion had an enormous impact on the Black Sea's ecosystems and commercial fish stocks. The loss of zooplankton allowed huge populations of phytoplankton to develop in a series of blooms that reduced the mean Secchi depth (the maximum depth to which a white disk lowered into the sea from a ship remains visible) from the normal average of 20 metres, to only five metres. Anchovy catches plummeted in 1990 to only 60,000 tons.

The situation in the Black Sea was mirrored by another environmental stress on its coasts. The economic decline of the Black Sea coastal countries and the political upheaval of transition to a market economy led to a lack of maintenance of waste treatment facilities for domestic sewage and industrial waste. Of course, many cities had never had effective sewage treatment but the general decline was evidenced by an increased frequency of outbreaks of waterborne diseases such as cholera and frequent beach closures due to unsanitary conditions. In Ukraine, for example, 44% of bathing water samples taken in 1995 did not meet the national microbiological standards.[8] This environmental problem, coupled with the decline in standards of tourism infrastructure and limited spending power of people in the region, also led to a sharp decline in tourist numbers and in the local economies.[9]

The state of the environment in the Black Sea in the early 1990s gave little reason for optimism. The economic crisis did however give some respite for pollution. Farmers were often unable to apply the quantity of fertilisers used in the former centrally planned economies. Many large energy-inefficient and polluting industries were forced to close. By 1996 there was already some evidence of recovery of benthic ecosystems on the north-western shelf of the Black Sea, albeit small. Furthermore, *Mnemiopsis* populations started to decline and the anchovy fisheries recovered, almost to their mid-1980s level. Most local economists and ecologists agree however, the pressure on the environment will return as the economies recover, unless urgent measures are taken to limit the environmental impact of renewed growth. Furthermore, new environmental pressures are emerging as a result of the rapid increase in the use of the Black Sea as a maritime transport route, particularly for the shipment of oil en route from the newly opened Caspian oil fields.

The response of people around the Black Sea to the deterioration in its ecosystem and to the environmental crisis around its shores needs to be considered at two levels - that of governments with or without the backing of the international community, and that of the local communities around the Black Sea itself.

International action for protecting the Black Sea

Perestroika sparked an openness which allowed more freedom to discuss environmental problems within the communist countries of central and eastern Europe. The spark kindled a fire which flickered and smouldered but never blazed. The change however, had profound consequences in the Black Sea region, and at the initiative of the USSR, the coastal countries (Bulgaria, Romania, Turkey and the USSR), met together in October 1986 to discuss a new convention to protect the Black Sea, along the lines of the Barcelona Convention in the Mediterranean, signed and ratified a decade earlier. Negotiations were slow because political events gradually overtook the negotiation process and with the break-up of the Soviet Union, Georgia and Ukraine joined the table as four countries became six. To their credit, the officials involved maintained a tenacious resolve to continue and the Convention to Protect the Black Sea Against Pollution (the Bucharest Convention) was signed in April 1992 and ratified by the six legislative assemblies within two years. Unfortunately it is still not being implemented.

In the meantime, another event had focused world attention on the need for new approaches to environmental protection. This was the 1992 Rio de Janeiro 'Earth Summit', the UN Conference on Environment and Development which approved the ambitious but under-funded 'Agenda 21', the agenda for sustainable development in the twenty-first century. In mid-1991, during one of the sessions of the working group charged with the task of developing Chapter 17 of Agenda 21 (the chapter on coastal and ocean areas), the idea emerged that the Black Sea might be one of the first regions to benefit from the application of some of the newer 'holistic' concepts of marine environmental protection. A mission to the countries of the region confirmed the interest of the governments of Black Sea countries in following this approach. It was evident that there was a strong need for a joint declaration of common policy goals - the Bucharest Convention provided a legal framework but it did not specify what was to be achieved and by when.

For its part, the United Nations Environment Programme agreed to host the negotiating sessions and work immediately started on consensus building. In April 1993, the Declaration was signed.[10] The resulting 'Odessa Declaration' is remarkable in two ways. The first is the spirit of consensus.[11] While negotiations were going on, Ukraine and Russia were engaged in a dangerous conflict about ownership of the Black Sea naval fleet and Georgia was suffering civil war. 'The environment has no political boundaries' explained Professor Sherbak, the then Minister for the Environment of Ukraine. The second remarkable feature was the commitment to a new approach to environmental policy making in the region, including much greater public participation and accountability. The 18 point Declaration also committed the countries to the development of a longer term 'Black Sea Action Plan' and boosted the interest and confidence of donors in supporting environmental protection in the region. The first political move had been made, and in the wake of the Earth Summit, these calls could not be ignored.

In June 1993, a three year Black Sea Environmental Programme (BSEP) was established with some US$ 14.3 million funding from the Global Environment Facility (GEF) and collateral donors. Shortly afterwards a Programme Coordinating Unit (PCU) was established to coordinate the activities of the BSEP. In order to share the task of programme implementation between countries, each Black Sea country agreed to host a BSEP 'Activity Centre', a specialist institution which addresses one aspect of the Black Sea environment.[12] The networks of institutions enabled specialists to 'reconnect' with each other and external funding provided additional training and modern equipment. Non-governmental organisations began to play a key role in the BSEP, holding national and regional fora. The programme also included major donors, as well as organisations such as the Black Sea Economic Cooperation, specialised UN agencies and international NGOs. Amongst many other things it has generated an urgent investment portfolio, implemented by the World Bank and instrumental in levering almost $100 millions of new investments with environmental benefits.

In June 1996, the BSEP demonstrated the potential of its institutional network by completing a systematic analysis of the causes of environmental degradation and appropriate measures for resolving them.[13] The task involved over one hundred specialists from more than thirty institutions plus many colleagues from the West. This study became the basis for the Strategic Action Plan. The analysis clearly demonstrates that the Black Sea environment can still be restored and protected. It proposed a series of actions to address the most serious problems. It also contains proposals for a revitalisation of the Black Sea economy in an environmentally friendly manner and highlights the need for a better regulatory framework and greater public involvement.

The Black Sea Strategic Action Plan[14] is a blueprint for the restoration and protection of the Black Sea. It provides a set of achievable concrete common measures[15] for the reduction and monitoring of pollution, fisheries management, the protection of natural ecosystems and sustainable human development. It also calls on the Danube River basin to reduce nutrient loads from agriculture, industry and domestic sources. The overall aim of the Plan is to *enable the population of the Black Sea region to enjoy a healthy living environment in both urban and rural areas, and to attain a biologically diverse Black Sea ecosystem with viable natural populations of higher organisms, including marine mammals and sturgeon, and which will support livelihoods based on sustainable activities such as fishing, aquaculture and tourism in all Black Sea countries.* Why marine mammals and sturgeon? Apart from humans, marine mammals and sturgeons are the highest life forms sustained by the Black Sea. Their presence is more than symbolic. Both depend upon a healthy, unpolluted and diverse Black Sea ecosystem. Dolphins depend on the healthy connection of the Black Sea with the world's oceans and sturgeon depend upon clean rivers for breeding as well as a clean Black Sea. The message is a clear one - the Black Sea offers opportunities for human development in co-existence with the most sensitive ecosystems.

Has the message of the Black Sea Action Plan been heeded? At the time of preparing this text, governments were busy preparing National Black Sea Strategic Action Plans, with detailed descriptions of how they are going to implement the larger regional plan. Despite this, there are still very few actions which have been implemented. Four years after ratification of the Bucharest Convention, there is no Secretariat in place to implement it, a situation that does not bode well for other measures which have been agreed. Certainly we need to appreciate that all six countries have faced, or are facing, huge economic and political difficulties (in five cases associated with the transition from centrally planned to market economies) and are under constant pressure from creditors to cut public spending and increase revenues. Pressing economic demands make it difficult to appreciate that the limits of economic growth are set by the sustainable use of the ecosystem.

The more we allow the environment to be destroyed, the less the scope for economic growth. If, for example, legislation for integrated coastal zone management is not agreed and imposed in many parts of the Black Sea, the coastline will become cluttered with a maze of ugly holiday homes, ports, factories and highways. The use of the sea as a repository for garbage or untreated sewage is evidence of a low societal value given to its protection. The beauty of the coast is already deteriorating in many parts and biological diversity is being lost as natural habitats are destroyed. Large stretches of the coast of Turkey, for example, are becoming dominated by rows of summer residences as people strive to own a key piece of the landscape but in so doing destroy it. Many tourists are already going elsewhere in search of a more attractive coastline and property prices are beginning to tumble. Both aesthetic and economic values are being lost. By careful planning and coexistence with nature, economic and aesthetic values can be assured and destructive impact minimised.

Environment still has a very low political profile in most countries of the world. Rather than incorporating environmental protection as an underlying principle of governance it is usually relegated into an 'environment sector' which itself often receives feeble support in the state budget. Officials in the Black Sea typically say, 'how can we support our institutions or programmes if we haven't received funds from the treasury?' The institutions which depend upon the Ministry of Environment for support often get too little too late and cannot fulfil their functions. Staff are paid salaries well below the cost of living and have to seek employment somewhere else in order to feed their families. Usually, the State Institution itself becomes the part-time job. This situation often applies to those professionals who are supposed to be providing essential data for environmental protection or the inspectors who are supposed to implement state and local legislation.

Local action for protecting the Black Sea

Until recently, in countries with a communist regime, individual or collective action on environmental issues was insignificant unless conducted within the limits defined by the government or through the channels of the communist party itself. In Turkey a stronger civil society is gradually emerging following difficult periods in the 1960s and 1970s.[16] In the entire region therefore, the 'environmental movement' is a new one. In a rather short period of time, non-governmental organisations have emerged, grown and gathered strength, though in many cases they fragmented and faded. One of their problems is that many NGOs were, and remain, small groups of specialists or enthusiasts, trying to raise funds for their projects and seeing the world through their own technical perspective. This function sometimes fills a vacuum left by weak government agencies. There are very few genuine community-based NGOs, however. Perhaps many people are still not fully aware of their democratic rights or are so preoccupied with the demands of survival in a transition economy that they have little spare time to seek other avenues to assert their rights.

The Black Sea has national NGO fora and a regional forum which have been the product of initiatives supported through outside donors. The NGOs themselves were mostly established as technical clubs rather than in response to deeply felt concerns amongst the general public. Consequently, there is little coherence between these groups, which is not to decry their usefulness and genuine intentions, but it does highlight the lack of initiative at local community level. Many donors have not understood this situation and see these NGOs as the only key to enhancing democratic processes, which is not necessarily true.

There are some recent examples of NGO action which are, however, remarkable and suggest that there is a limit to the apparent apathy shown by some local communities. In the city of Zonguldak in Turkey, a local community-based NGO, predominantly organised by local women, successfully took the Government to court over the direct discharge of fly-ash to the Black Sea from the local thermoelectric power station. The station has now been equipped with the necessary treatment facilities. Near the Black Sea port of Novorossiysk, the construction of an oil terminal is planned. The Black Sea is rapidly becoming an oil superhighway, driven by the unabated demands for energy from western markets - a new cause for concern for an ecosystem which is already sick. Partly as a result of efforts by the local NGO Aquatoria, the local public was mobilised and, in a public hearing, rejected the results of an official environmental impact assessment which supported the construction on a site considered to be of great natural beauty. The stakes are very high as the terminal is a key element of Russia's oil export plans.

One element that may be missing in the efforts to find solutions is the current lack of emphasis on supporting and strengthening local communities. Well-meaning donors try to build the capacity of central government agencies or to support large national NGOs, assuming that they are similar to the movements in the West. This approach by itself is insufficient as central governments and other centralised organisations are not adept at transferring the benefits of technical assistance from capital cities to the regions and fully involving them in the decision-making process. Until recently, many local authorities remained unaware of the $20 millions already spent in international efforts to protect the Black Sea and may feel they play a role as observers rather than full participants in guiding the overall effort. The current development of National Black Sea Action Plans involving key sectors at the local level may improve this situation.

Different sectors in local settlements need to be brought together if a community spirit is to exist at all. In this respect, the institution of 'International Black Sea Day', annually celebrated throughout the Black Sea on 31 October, has been important in prompting the involvement of local authorities, donors, NGOs, business groups, schools and the media. The initiative focuses the often disperse efforts of these groups on a single common objective. The fact that practical activities are being undertaken in many towns in six countries on the same day excites the imagination and gives a sense of international solidarity which the media is quick to grasp. Unfortunately, incentives for individual or collective environmental action at the local level are virtually absent. These would help[17] to stimulate citizen initiatives and give a higher status to environmental protection in society.

If there is to be an effective improvement in environmental protection at the local level, the needs of each community require thoughtful examination so that actions can be proposed which are sensitive to the culture and which take account of the social and economic constraints of local society. Strategies such as 'Local Agenda 21' have been readily embraced and adapted by some towns but it is important not to impose a model from a totally different western culture and assume that it will work. Support is certainly needed to train all sectors of the community in environmental awareness, as well as to offer ideas for practical protective and remedial measures. These measures can be very simple; for example the provision of receptacles for used car oil, combined with public information and fines for those who throw their oil down the drain, can significantly reduce river and sea pollution and improve the efficiency of sewage treatment. Many citizens are aware of the problems, at least superficially, but expect someone else to resolve them. In some towns, the appearance of a local religious leader on television talking about the state of the environment in the community and the responsibility of each citizen, might make a significant difference at local level, especially if practical follow-up actions are proposed and taken. Protection of the environment is an issue around which a genuine community spirit could develop.

Beyond science and politics: caring for the Black Sea

Technically speaking, there is considerable evidence that the Black Sea's decline was quite sudden and intrinsically linked to the Green Revolution in agriculture in which higher levels of production were sustained by the use of chemical fertilisers and pesticides. Before simplistically denouncing the Green Revolution as an irresponsible action, it is worth recalling that it provided a relief from famine for millions of people in the developing world, albeit in some cases a temporary relief. In many respects however, the Green Revolution has proven to be another example of the 'technological quick fix' where a problem is solved by applying the best available technology without considering its wider systematic implications. This is easily said in hindsight of course but, at the time, the vision of new technology to revolutionise food production was accompanied by euphoria and there had been few examples of the massive destruction of ecosystems comparable with those we are seeing today.

The Black Sea basin is full of evidence of the application of the technological quick fix. Industrial quotas were met at a heavy cost to the environment since the environmental laws, on paper extremely strict, were overridden through a 'law of exception' which set the production of strategic industries above protecting the environment. Agricultural quotas were met (or often not met) by massive use of chemical fertilisers and pesticides, the application of which was sometimes made without consulting the farm managers. Huge production units for animals spilled untreated waste into rivers. Massive irrigation schemes denied vital water supplies to rivers and lakes, causing, for example, a radical change in the circulation of the Sea of Azov which virtually eliminated the stock of wild sturgeon.

Paradoxically, the centrally planned system also brought some environmental benefits. The system of National Parks and protected areas in the Soviet Union was one of the strongest in the world and these areas benefited from considerable state subsidies and rigorously enforced regulations. The management of many of these parks has now broken down as they have not found their place in the 'free market' and receive little or no state support. Furthermore local people have been segregated from these areas for years with little or no explanation regarding the real need for nature protection. The staff of the parks have become disillusioned and frustrated as they see that the *nature* which they protected is now regarded as *natural resources* to be mined or plundered. There are exceptions, like the remarkable Danube Delta Biosphere Reserve, visited by the symposium, but these exceptions are few in number.

The withholding of information has been (and often remains) a much-used tool for the maintenance of power by totalitarian regimes. Information is withheld because knowledge of the true situation might result in dissent. Absence of accurate information engenders a reliance on rumours and anecdotal information or, in many cases, fuels apathy. Freedom of information, a precept of many western societies, has not yet been fully embraced in post-communist cultures. The absence of reliable information from the past makes it difficult to understand trends or to provide a basis for environmental management decisions, but programmes such as that of the Black Sea are rapidly changing this situation.

With a lack of accurate information, blame is often apportioned inappropriately. The decline of fisheries in the Black Sea led to an angry exchange of blame: Chernobyl, dolphins, Turkish fishermen and Austrian farmers have all been cited in local newspapers as 'the cause'.

This only points to the need for clear public information on the issue. As it happens, scientists have been gathering data for some time on the inputs of dissolved nitrogen and phosphorus compounds to the Black Sea (not an easy task). To the best of their knowledge,[18] some 14% of total nitrogen is from Bulgaria, 27% from Romania, 12% from Ukraine, 10% from the Russia Federation, less than 1% from Georgia, 6% from Turkey and about 30% from the non-coastal countries (Austria, Belarus, Bosnia and Hertzegovina, Croatia, Czech Republic, Former Yugoslavia, Germany, Hungary, Moldova, Slovakia, Slovenia). In the case of phosphorus, the figures are Bulgaria 5%; Romania 23%; Ukraine 20%; Russia 13%; Georgia 1%; Turkey 12% and 26% for the remaining countries, a similar story to that of nitrogen. The importance of showing these numbers is to illustrate that nobody is 'innocent', not even the Georgians whose low percentage input reflects the current collapse in the coastal economy, probably a temporary feature.

Incidentally, the reader might wonder why the contribution of Romania is so high. Romania's entire territory drains into the Black Sea, mostly through the Danube. The industrial and agricultural practices adopted during the former political regime paid little regard to environmental protection, especially in the Green Revolution. Now that the economy of Romania is market-based, many subsidies on fertilisers have been removed and large animal production complexes are closing. The decrease in fertiliser use is beneficial to the environment but unless alternative and cost-effective agricultural practices are adopted, there will be enormous social problems of unemployed farm workers unable to compete with cheap food exports from places where cheaper production techniques are applied. A similar situation prevails in neighbouring Moldova where large animal complexes have also closed but where smallholders now have excessive numbers of animals literally in their back gardens, in very unsanitary conditions. Human health is already declining in these places and shallow wells, the main local water supplies, are polluted. There are no simplistic solutions to these problems unless consumption patterns themselves are changed - and how can countries with rampant over-consumption in the West demand changes of their poorer neighbours in the East? The renowned ecologist, Ramon Margalef commented that 'the job of the technical human in the preservation of some properties of our biosphere and the conservative management of the environment might come simultaneously with, and perhaps only second to, the need to develop forms of behaviour, within our species, less conducive to the now free running mechanisms of developing and maintaining inequality'.[19]

Understanding the root causes of environmental degradation in the Black Sea and doing something to redress them is a challenge that faces anyone who cares about the Black Sea. Terms such as 'environmental crisis' hide the reality that the problems are of human origin and the solution is in our hands. The technological quick fix represents faith in the power of humanity to dominate nature, to understand the workings of all things and to be able to manipulate them at will. Modern science has no such pretences and the probabilistic concept of *uncertainty* is a measure of just how categorical we can allow ourselves to be in our assertions. We attach great certainty to some observations of some physical phenomena but a much lower degree of certainty on the interrelations within the web of life. This is the basis of concepts such as the *precautionary approach* which accepts the current limitations of science and requires action to be taken when there is reason to suspect deleterious environmental effects of human activities, irrespective of the existence of conclusive scientific evidence.

In the Black Sea region, as with many other places in the world, people are becoming increasingly disconnected from the environment which supports them and find it difficult to relate their individual or collective actions to the visible deterioration in their own habitats. Large sections of the population feel confused by the rapid changes in the society around them, are apathetic or deeply suspicious of the actions of governmental authorities. This apathy towards individual social responsibilities is not entirely new but, in former communist times many people genuinely believed in the system, irrespective of what western society may regard as 'indoctrination' or an absence of individual freedom or democratic rights. For many of them, the system offered a feeling of personal security and provided a sufficiency of daily needs, education, health services and employment opportunities. There is no doubt that this system was economically (and politically) unsustainable and was achieved at a serious cost to the environment and was often repressive, but the person on the street was generally unaware of this situation and often remains so.

This, of course is one of the realities ignored by many western visitors, consultants and the like who have imagined themselves as 'liberators' arriving to the welcoming arms of dejected Soviet citizens. Quite the contrary, the new situation of many former Soviet or Eastern European citizens is one of great economic and psychological instability; some have been able to grasp the opportunity and 'rise' in the new social context, others, the majority, find the transition deeply disturbing and do not have the tools to adapt. Western support has focused on rebuilding the economy, opening the door to foreign investors and upgrading basic services. This is, of course, very important but much of the new economy is highly extractive, capitalising heavily on 'natural resources' in the hope of 'kick starting' productive industry (at least this is an excuse often quoted). Forests are being cut, oil and gas lines built and planeloads of hunters arrive to pursue their 'sport' from the comfort of a helicopter. Many of the investments, especially those negotiated bilaterally, feed off the continuing belief in the power of technology to solve the problems 'at the end of the pipe'. The technology itself is often inappropriate. The case comes to mind of a municipal garbage incinerator in a Black Sea coastal resort built by a foreign company using 'best available technology from western Europe'. It stands idle, two years after completion. Prior to construction, the manufacturers failed to study the different composition of local garbage from that of their own country. The incinerator was unable to reach the high temperatures needed to operate safely in an environmentally acceptable manner. Garbage with such a high organic content could have been separated and partly composted, but such techniques require new consumer practices, new attitudes by the authorities and do not result in such lucrative deals with foreign companies.

Perhaps the writer has painted an image of despair. What is being portrayed might be described as a loss of faith. Substituting one blind faith with another is a formula for spiritual disaster if not accompanied by deep reflection or enlightenment. In this process, there is a need for the individual to reassess and reassert his or her values. Much of the present book is about the reassertion of values though it may not seem immediately obvious when viewing the contents page. Environmental values which offer hope for the future integrity of our ecosystem reject the utilitarian *anthropocentric* vision of it as something to be exploited relentlessly for human benefit, in favour of a vision of cherishing nature for its own sake, recognising the need for humanity to make mindful use of nature without destroying its complexity and beauty.

Greater environmental awareness is certainly one key to changing basic values, but it is naïve to regard awareness campaigns as being sufficient. They have to be accompanied by an understanding that there are viable options for sustaining the natural environment and human welfare. Some of those options require technological solutions. It also has to be understood that the situation requires a personal and collective commitment by all individuals in society. Understanding has to be followed by action. This is easier said than done in a society in which people are barely able to cover the material needs of their families, spend inordinate amounts of time confronting dehumanising bureaucratic obstacles, and often feel that individual action will be ignored or repressed.

The message of *awareness-choice-commitment-action* is one which finds most fertile ground in younger generations. The importance of environmental education should not be understated - not just formal education in the classroom but bringing children in direct contact with nature in a manner which inspires them and promotes deep feelings as well as knowledge. Children need to become aware that the litter in their parks, streets and beaches is just as important a source of environmental degradation as the factories belching fumes and toxic wastes which they may only read about in books or see on the television. This example is not trivial, values which discourage littering can also be the foundation of changing attitudes and behaviour towards more complex and difficult to understand issues such as recycling, making tourism compatible with environmental protection or eliminating waste discharges from industry or agriculture. Recently, a group of teachers from all the Black Sea countries came together to share experiences and to plan practical actions for improving environmental education in their schools. Their work is severely hampered by a lack of funding. Raising funds for environmental education is very difficult. We are reticent in recognising that much of our foreign aid is focused on improving trade and little is available for fostering the transfer of knowledge and wisdom to a new generation of citizens who will have to tackle the legacy of our current very unsustainable lifestyles.

The Black Sea symposium made an important contribution towards sharing individual feelings about the environment as well as technical knowledge. It recognised that the wisdom needed for protecting the environment involves the spirituality of human contact with nature as well as scientific knowledge. It reminded us, each from our own perspective, that there is no room for arrogance: we are part of the complexity of nature and not its ruler. The crisis of values in the Black Sea is part of a planetary crisis which manifests itself in many different ways under the circumstances of every individual and every community. Spirituality lies deeply within every human being, sometimes unrecognised. If we allow ourselves spiritual contact with nature, we are surely bound to love and protect it. Few of us can stroll along the Black Sea coast on a summer evening, bathed by the glow of sunset and the fragrance of seaweed and magnolias, without pausing to gaze out to sea at the setting sun - not out of curiosity but with a deep sense of awe that can be the gateway to spirituality.

Notes

1. The current review is largely based upon the work of the Black Sea Environmental Programme, an initiative sponsored by the Global Environment Facility and major donors, notably the European Union (Phare and Tacis). Detailed information on the development of the GEF Black Sea Environmental Programme may be found in its Annual Reports (for 1994 - 1997) and its regular newsletters. These may be obtained from the BSEP Project Implementation Unit, Harekat Köskü II, Dolmabahce Sarayi, 80680 Besiktas, Istanbul, Turkey.

2. BSEP (1997) Black Sea Transboundary Diagnostic Analysis, (ed. L.D. Mee) United Nations Publications, New York. ISBN 92-1-126075-2, August 1997.

3. Zaitsev, Y. (1992) Recent changes in the trophic structure of the Black Sea. Fish. Oceanogr., 1(2): 180-189.

4. Kos'yan, R.D., Magoon O.T. (eds) (1993) Coastlines of the Black Sea. Proceedings of the 8th Symposium on Coastal and Ocean Management, Coastal Zone '93. Coastlines of the World, American Society of Civil Engineers.

5. Mee, L.D. (1992) The Black Sea in crisis: The need for concerted international action. Ambio 21(4): 278-286pp.

6. MacLennan, D.N., Yasuda T., Mee, L.D. (1997) Analysis of the Black Sea Fishery Fleet and landings. Black Sea Environmental Programme, Istanbul.

7. GESAMP (IMO/FAO/UNESCO-IOC/WMO/WHO/IAEA/UN/UNEP Joint Group of Experts on the Scientific Aspects of Marine Environmental Protection) (1997). Opportunistic settlers and the problem of the ctenophore *Mnemiopsis leiydi* invasion in the Black Sea. Rep.Stud.GESAMP, (58):84p.

8. op. cit. 2.

9. BSEP (1996) Black Sea Sustainable Tourism Initiative (Background report), Istanbul, Turkey.

10. Hey, E., Mee, L.D.(1993) Black Sea: The Ministerial Declaration, an important step, Environ. Policy and Law, 23(5) 215-220.

11. The author Neal Ascherson was present during the signature of the Odessa Declaration and relates the remarkable spirit of consensus in his book *Black Sea*.

12. Emergency Response to Oil Spills (Varna, Bulgaria); Fisheries (Constanta, Romania); Pollution Assessment (Odessa, Ukraine); Coastal Zone Management (Krasnodar, Russia); Biodiversity (Batumi, Georgia); and Pollution Control (Istanbul, Turkey). The BSEP also created a Black Sea Data System and a Black Sea Geographic Information System.

13. op. cit. 2. BSEP publications include a Technical Series, published by UN Publications, New York and currently numbering eight volumes, a Black Sea Studies series, focusing on economic and social information and newsletters and books for the wider public.

14. BSEP (1996b) Strategic Action Plan for the Rehabilitation and Protection of the Black Sea, Istanbul, Turkey, (available as per ref. 1). The Action Plan was formulated by the BSEP National Coordinators (mostly Deputy Ministers of the Environment) and their advisors and signed in a Ministerial Conference in Istanbul on October 31, 1996.

15. See also the final section, Recommendations for Helping to Protect the Black Sea.

16. Pope, Nicole and Hugh (1977) *Turkey Unveiled*, John Murray, London.

17. This matter is taken up in more detail in the Recommendations.

18 Topping, G., Sarikaya H., Mee L.D. (1998) Sources of pollution to the Black Sea. In: Mee, L.D. and Topping G. (eds) (*in press*) Black Sea Pollution Assessment. UN Publications, New York.

19. Margalef, R. (1997) Our Biosphere. *Excellence in Ecology* 10 (ed. O. Kinne), Ecological Institute, Oldendorf.

Chapter 1

The Challenge of the Black Sea

Address of His All Holiness the Ecumenical Patriarch Bartholomew I

Among many prevailing misconceptions is the idea that religion in general, and Christianity in particular, are interested only in life after death, as if by struggling to divest one's self of a material body, one might be changed into a liberated spirit, as Plato taught. However, leaving the arguments of other religious teachings and philosophical schools to their own adherents, we will focus on the Eastern Orthodox Church, in the knowledge of our responsibility.

It is a qualitative element in our faith that we believe and accept that the Creator Who loves us fashioned humankind in its psychosomatic reality and also fashioned the material universe according to His image and likeness, making it and us 'very good'. Consequently, we stand far afield from any dialectical opposition of matter and spirit, body and soul, or the present world and metaphysical reality. We reject this Manichean perception and live in harmony with that which we are, in harmony with the material and spiritual world in which we live and move and have our being. We live in harmony with God, the creator of all things and with all things which He has created.

Obviously, we recognise a certain hierarchy in the created order and we accept a gradation of proportionate value in the harmony of creation, inasmuch as the all-wise Creator has arranged all of creation in the fullness of His love. Even when we cannot know the specific reason of creation and the existence of an objective reality or even the property of a physical law, we can be absolutely certain of one supreme Logos, Who has established and apportioned all things with immanent rationality. Because of this we hesitate to opine that human remedies which interfere with the natural order are better than those found in the natural conditions provided by God. However, it does not follow from this that we consider the natural order to be an intangible 'taboo,' for we possess the commandment of God that we should have dominion over the earth.

We should focus our attention from the outset on the proper limits of our interventions, such as they are, and around a consideration of what might be considered appropriate interventions. We can speak in a positive manner and without hesitation about these things.

In describing the appropriate limits of our interventions in the natural condition of the material world, in each particular instance we stand on a razor's edge, which itself is a measure of our own spiritual condition. Our answer presupposes a careful weighing of innumerable factors, which must take place with prudence and dispassionate discretion; this unfortunately is rare, much like our own virtue is rare.

The first part of the word 'ecology', is derived from the Greek word 'oikos' meaning our 'home'. There is more to our 'home' than people, land, water, plants and animals - there is the Spirit. Yet people rarely take time to look at the beauty of this world or to feel the spiritual connection between us and our surroundings. In order to change our behaviour towards the 'home' we all share, we must rediscover spiritual linkages that may have been lost, and reassert human values in our behaviour.

And so we stand before nature, which has in recent times come to be called the 'environment' with the same affirmation of its own worth, as that with which the Creator of all things characterised it as 'very good'. But at the same time, we face the reality of the hierarchy inherent in the design of the cosmic plenitude, both the material and the spiritual, and we assign it to its proper place. We do not make of nature an absolute value, nor do we deny it to the point of absolute negation. Nature exists for the service of humanity according to the material portion of our nature; it is not humanity that exists for the sake of nature.

With this understanding, the core of our concerns for the Black Sea and the environmental dangers threatening it is focused on those people who live in the surrounding regions of the Black Sea, and not on the Black Sea in itself. In as much as we are interested in preventing a catastrophe of the Black Sea as a biotope, we declare that our most vivid concern is for our fellow human beings who dwell either near or far from it and who may not be aware of the danger that threatens their life, health and livelihood through the disruption of the natural ecological balance of the Black Sea.

The Black Sea is a remarkable place with a long human history. Unlike other seas on our planet, it was formed comparatively recently, when the waters of the Mediterranean broke through the Bosphorus valley, flooding a vast low-lying freshwater lake, some seven or eight thousand years ago.

People living in primitive settlements will have watched this great event as a salty cascade surged down the valley to give new life to the new sea. Subsequent generations have admired the beauty of this sea, enjoyed the fish it provided, and sailed its waters. Different civilisations have come and gone, but always the sea seemed an immutable part of eternity, governed by the 'everlasting covenant between God and every living creature upon earth' that was proclaimed in the book of Genesis after Noah and his Ark survived the mighty floodwaters. Now, almost as dramatically as it was formed, the people see change, fish are disappearing, crystalline waters are becoming choked with green algae, the beaches and harbours are badly polluted, and the Black Sea's beauty is soiled by the hand of mankind.

We are deeply concerned for their daily life, for it too is a gift of God, 'very good' indeed and destined to be used as a starting-point for an eternal life, continuing beyond the grave into the infinite, when the bodily substance of our nature is transmuted into a spiritual reality.

To ensure the furtherance of natural life, we consider ascetic self-denial to be necessary, and the reduction of many material pleasures to be beneficial. However, such deprivation should certainly be a voluntary and conscious decision, and not something imposed by external constraints and consequently without ethical merit. So from this perspective we join forces with all those who are concerned for the self-sufficiency of those material goods necessary for survival, but we do not approve of overproduction, irrational exploitation and greed. We seek and urge the Patristic and Aristotelian sense of moderation and balance and would dissuade you as much from forming an idol of wealth as from reducing wealth to an evil in itself. For wealth is the accumulation of the abundance of human labour and natural resources. Wealth is both valued and reviled in an ethical sense according to the manner of its concentration and the manner of its use.

It is obvious that the rapacious exploitation of natural resources, which derives from greed and not from an extraordinary circumstance of need, entails a debilitation of nature to the extent that it cannot renew its own productive powers. Such exploitation is as unethical as it is irrational, given that, as is observed sociologically, a lack of ethics coincides with a lack of intelligence to an astounding degree, because someone who is truly intelligent understands the injurious nature of bad character. Again, it is equally evident that the utilisation of concentrated wealth for selfish purposes, for one end or another, for good or bad, is ethically disdainful. Rather the utilisation of the concentrated surplus of human effort and natural resources forms a basis of the furtherance of the means of production, aimed at the improvement of the quality of life in human society and for the allocation of various aspects of education, public health and other possibilities for the development and enrichment of the spiritual life. This justifies wealth as a shared property, expended for the common good, even if it is managed by only one part of human society.

The Orthodox point of view and teaching is that the environment was created so that we might evolve and develop within it in body and soul. In addition, we ought to preserve the divine law and do those things that are useful and reasonable for our self-sufficiency and not for any kind of hoarding. If there should happen to be an excess without draining the vigour of nature dry, this ought to be employed again for the common good. Thus, religion propounds eternal truths and maintains a sensitivity to all members of society because of the danger of falling away from those same truths.

It is at this point that the work of science begins, for science certifies the situation, proposes measures that would delay the worsening of this situation, and organises the effective application of these measures. Not infrequently, whenever science considers it possible through the measures at hand, it also proposes the transformation of the destroyed elements in the environment.

This symposium has the blessing of gathering together, as its title declares, representatives of religion and science. All of you enjoy the blessings of the Holy and Orthodox Great Church of Christ, and personally speaking, most certainly the love and the prayers of our Modesty; you enjoy the patronage of the President of the European Commission, His Excellency Jacques Santer. The subject of this symposium is of interest to many peoples and their governments. The circumnavigation of all those interested countries as the venue for this symposium constitutes a conscious decision, aimed at sensitising individuals and whole peoples to the significance of the biotope of the marine region of the Black Sea for their life. Environmental change and destruction on such a vast sea-wide scale, through the thoughtless contribution of many, wreaks vengeance over a long period of time and in all directions. The redress of such circumstances is a most onerous task and requires the cooperation of all concerned parties; that is why it constitutes a challenge to the ethical and social conscience of us all and substantiates a criterion of the degree of our spiritual condition.

The Black Sea is spoiled and its biotope destroyed as a result of the indifference of countless people, each of whom contributes to a greater or lesser extent to the unfortunate result. A ship's captain, for example, protected by the fact that it is not easy to prove that he discards his ship's polluted waste into the sea, is as responsible as the person who lives far from the coastline, but who discards his own destructive waste into the river which finishes up in the Black Sea. Unless everyone is made sensitive to the harmful nature of such actions, it is almost impossible for any endeavour to improve the situation to succeed.

It is precisely here that the fundamental role of religion is revealed. Religion can inspire the behaviour of every individual; here also is the role of individuals and mass movements, who can spread the necessity and benefit of these behaviours.

The fact is that the negative results springs from the contribution of many people who may hardly be aware of their own responsibility, because their contribution to the overall result is relatively minor. If there is going to be an overturning of this problem, it will require the cooperation of everybody; in other words, it will reveal the necessity of achieving a social ethic and a common effort. There is no single deed which benefits others, without it first and foremost benefiting ourselves.

This sense of a common fate, the polar opposite of the widespread individualistic and self-interested perception which is short-sighted in its appreciation of the world, is a basic teaching of the altruistic Christian faith, and especially of Orthodoxy. Without it, facing problems such as those of the Black Sea would be impossible.

Surely there are not many today who are ignorant of the influence of prevailing ideas and beliefs on our lives. To bring new ideas to predominance we must focus on the acquisition of adherents and on the spreading of ideas. With this understanding, we seek also through this symposium to activate for the common good the dormant sense of responsibility which exists among every people and in each individual. Thus can the aim of a healthier Black Sea be more easily achieved for the benefit of all mankind.

We are deeply pained by the suffering which, as a result of human indifference, keeps piling on our fellow human beings. We call on every conscience to awaken! We invite those of our beloved fellow participants whose consciences are already sensitised to a virtual apostolic commission, to disseminate ideas about the necessity for a common confrontation of these problems.

We would like to express our thanks, first, to all of you for being here, each for their own personal reasons, and thus showing a commitment to a cause which is dear to our heart. Thanks to all whose organisational skills have made this symposium possible. Thanks to all those who have contributed, whether with resources, time, effort, or in other ways, and above all, thanks to God for bringing us together in this place and at this time so crucial to the future of the Black Sea in particular and mankind in general.

Finally, we pray wholeheartedly for the success of the proceedings of this symposium. We welcome all the participants and thank them for their kind response and attendance. We pray for the health, joy and inspiration of each of you personally, and eagerly await to hear of your detailed and informed proposals.

The grace of God be with you all.

A Common Responsibility

Ritt Bjerregaard

This symposium on the Black Sea is an historical event in more than one sense. It was at Trabzon that the 10,000 Hellenic soldiers, led by Xenophon 400 years before Christ, saw the sea again after years of warfare and hardship and cried out the famous words which have resounded in the minds of people up through the centuries: 'Thalassa - thalassa!' - 'The Sea - the sea!'

Throughout history the sea has retained its power over the minds of men. It has divided peoples and cultures - and brought them together. People have crossed the sea in order to trade and to explore new worlds, but also to conquer and plunder. The sea has opened up the world to us. Cultural influences - from the most mundane products to the loftiest ideas - have come by boat from all corners of the world.

In our minds and imaginations the sea has retained its character of wilderness, of 'real' nature, untamed by man. This perception is misleading. Even if previous human efforts to 'rule the waves' have failed, we now know that there is a real risk that we might end up destroying the sea because of the pollution of rivers flowing into it and direct discharges of wastes into the sea.

Today, 24 centuries after Xenophon's soldiers first saw the Black Sea at Trabzon, we can see the whole of our planet Earth, photographed by satellites in space. The picture of this small blue planet, largely covered by water, suspended in infinite space, which we have all seen on television, presents us at a glance with the inescapable challenge facing us. We are confined to this small planet which we all share - and which we will also have to share with future generations.

It brings home to us the moral responsibility we have, to ensure that the social and economic development we seek for a still increasing number of people does not happen at the expense of the basic conditions for future life on this planet. This is what we mean by the necessity of ensuring sustainable development. This is the vision which came out of the World Commission in the Brundtland report on 'Our Common Future'. And this is the vision which was confirmed in the declarations adopted at the Earth Summit in Rio de Janeiro in 1992.

We have already come a long way since the United Nations Conference on the Human Environment in Stockholm in June 1972, which first drew political attention to what was happening to the environment. However, let us be quite clear: there is no reason for complacency. Seen in the light of what is needed in order to achieve sustainable development, we are only at the very beginning. Much more remains to be done. The burden on the environment is increasing: production and consumption in the developed countries still use more and more resources; human activities influence and change our climate; 40% of the world's population is facing water shortages; forests and biodiversity are disappearing at an alarming rate.

These facts are not only known by scientists and governments, but widely recognised by the general public in Europe. Public opinion polls continuously show that ordinary people all over Europe recognise that there is a need for strong action to protect the environment. They recognise the need for new types of determined action to achieve this in their own countries and they are ready to pay the price. They are also ready to assist developing countries in their struggle towards sustainability.

Today the European Union is in the lead in the international negotiations to tackle some of the major threats to our global environment, such as depletion of the ozone layer and reduction of greenhouse gases, which are gradually causing climate change. However, efforts to achieve solutions to global environmental problems cannot be limited to global treaties and conventions. We must start at home, setting our own house in order, and we must assist our neighbours in their efforts to do the same.

The European Union is built on the fundamental idea of ensuring peace, democracy and prosperity through the establishment of a common market as the driver for a process of continuous political and economic integration between countries which had previously been at war for centuries.

In 1992 the pursuit of sustainability was introduced in the Treaty of the European Union as one of its fundamental principles. At the Summit in Amsterdam in 1997, the European Heads of State confirmed this commitment and specified that sustainable development should be a guiding principle for all sectors of policy of the Union.

The European Union has grown from economic cooperation and integration between six countries to a fully-fledged political and monetary union comprising 15 European member states today. The next round of enlargement, with countries from Central and Eastern Europe, is already in preparation.

Ever since the great events that unfolded around these parts after 1989, Europe has been seeking to come closer to the peoples separated from us for so long by artificial barriers. Association Agreements, opening up a pre-accession process, have been signed with ten Central and Eastern European countries, including Romania and Bulgaria. In December 1997 the European Heads of State discussed the strategy for enlargement proposed by the European Commission.

But the European Union's interest in the development in the East does not stop at the borders of the prospective new Member States. The European Union has established

Partnership and Cooperation Agreements with Russia, Ukraine, Moldova, Georgia, Armenia and Azerbaijan. A customs union with Turkey rounds off a series of important agreements that politically and economically bind the countries of the Black Sea area to the European Union countries.

We in the European Union are increasingly aware of the need to save the Black Sea. We also accept a common responsibility, knowing that some of the problems originate from our region via the Danube River. The Danube provides a direct link between the Union Member States and the Black Sea. This is why the Union is also a signatory to the Danube Convention, providing important assistance for the improvement of the Danube and ultimately of the Black Sea.

The setting up of frameworks of cooperation with the European Union has led to concrete forms of assistance provided in multiple forms through many Union programmes. This European assistance has extended into all areas of the economies of the countries concerned and has contributed considerably to the establishment of a number of institutions. In the Black Sea area European Union aid has helped establish, among other things, the Black Sea Environmental Programme which has led to the Convention on the Protection of the Black Sea and its secretariat in Istanbul.

In fact the countries of the region have made great progress in the institutional development of their cooperation in the area. The Black Sea Economic Cooperation and its institutions are firmly established and are ready to make their contribution to the economic development of the region.

When we turn to the future we see that many of the building blocks are in place for an accelerated development of the countries of the region through the appropriate use of their natural resources. Their privileged location in the confluence of two great continents allows them to play a key role in the further development of the transportation and energy networks that will be the backbone for their development.

The realisation that we are menaced today by great global and regional environmental degradation places on these countries a special obligation to develop all their resources in a sustainable manner.

I grew up on the shores of another landlocked sea, the Baltic. When I was a child the Baltic Sea was a military border between East and West. Today most of the countries surrounding the Baltic Sea are - or are about to become - members of the European Union. Today it is a realistic description - and not just an expression of Cold War propaganda - to call the Baltic a 'Sea of Peace'. Even so, the Baltic is also a sea in crisis. Like the Black Sea, it is threatened by pollution and eutrophication.

The Black Sea is about to become a sea on the borders of Member States of the European

Union. It is still a sea in crisis in more ways than one. We all have a responsibility to work for a future in which the Black Sea will regain its equilibrium and become a sea bringing people and cultures together, not dividing them.

A satellite picture gives us the right perspective on our world: we cannot see the national or military borders, which have cost so much blood and suffering. We can see that we live in one environment, the planet Earth. And like Xenophon's soldiers we can see the Sea!

Developing the Common Ground: the Purpose of the Symposium

Jane Lubchenco

The Black Sea is a microcosm of Planet Earth. Its environmental problems are multiple and complex. They are the result of a broad spectrum of activities occurring locally, far upstream in the air- and watersheds, and even globally. This symposium focused on the Black Sea because solutions to its problems are urgently needed and because pathways to those solutions may enlighten, inform and inspire the global challenges.

The role of science in identifying and implementing these solutions is to provide information. Documentation of the changes that have occurred, understanding of the causes of different changes and analysis of the likely consequences of different possible actions will enable more informed decisions. Science alone is insufficient.

The unusual approach of bringing together scientific and religious perspectives to address environmental issues was reflected in *Symposium I, Revelation and the Environment* (Hobson and Lubchenco, 1997). Symposium I began the task of redefining humanity's relation to and responsibility for the environment. Symposium participants discovered common ground in their concern for the global trajectory and focused on the need for more responsible stewardship. They found energy in this common ground and a strong desire to build on that promising beginning. The goal of Symposium II is to take the next step - to work and learn together in order to share knowledge and to effect change. Building on the first symposium's global-to-local perspectives, this paper first summarises some of the most important global-scale environmental issues, then considers specific environmental problems of the Black Sea.

Human actions are altering the Earth in ways that are serious, startling and staggering. Scientists have recently documented the following global-scale changes (Vitousek and Mooney et al., 1997):

- Between one-third and one-half of the land surface of the planet has been transformed by human action (Vitousek et al., 1986, Kates et al., 1990, Daily, 1995);
- The carbon dioxide concentrations in the atmosphere have increased by nearly 30% since the beginning of the Industrial Revolution (Schimel et al., 1994);
- More atmospheric nitrogen is fixed by humanity than by all natural terrestrial sources combined (Vitousek and Aber et al.. 1997);
- More than half of all accessible surface fresh water is put to use by humanity (Postel et al., 1996);

- About a quarter of the bird species on Earth has been driven to extinction (Olson, 1989, Steadman, 1995);
- Two-thirds of marine fisheries are fully exploited, overexploited or depleted (Botsford et al., 1997, FAO 1994).

The conclusion from this summary is inescapable: we live on a human-dominated planet. Our actions are causing rapid, novel and substantial changes to Earth's ecosystems and hence to people. These changes are happening faster than ever before; they cover a larger area than ever before; and many of them represent new kinds of changes - for example, the development of synthetic organic compounds (Lubchenco, 1998). We live in a fundamentally different world than has ever existed on Earth, a world transformed by human activities. Moreover, the world is changing at ever increasing rates - and faster than our abilities to understand the changes. The rate and extent of human alteration of the Earth must thus affect the way we think about the Earth, the way we think about our responsibilities to each other and to all of life on Earth.

When asked how we depend upon nature, most people will focus on the goods, the things we derive from nature: for example, food, fibre, medicines and genes. We are learning the hard way that we also derive essential services from nature. These services are provided by intact, functioning ecosystems. They include regulation of climate, control of floods, provision of fertile soil, purification of air and water, and provision of pollinators (Daily, 1997, Chichilnisky and Heal, 1998). We take most of these ecosystem services for granted. They have always been there, free of charge. We begin to appreciate them only when we lose them through disruption of the ecosystems providing the benefits.

Together with the goods, these ecosystem services provide the life support systems for all of life on Earth. Because we are dependent upon both the goods and the services, and because they are provided by intact functioning ecosystems, it behoves us to protect these ecosystems - the forests, coral reefs, seagrass beds, open seas, riparian zones, grasslands of the world. Our health and well-being depend upon these systems.

The global-scale transformation cited above and the consequent disruption of the provision of goods and services have immediate and long-term consequences for people. As we begin to appreciate the myriad and complex ways in which humanity depends on the environment, we are redefining what 'the environment' means. Specifically, the definition of environment is being expanded to include human health, the economy, social justice and national security (Lubchenco, 1998). All of these are environmental issues because human well-being, prosperity, justice and security depend upon intact functioning ecosystems.

Environmental degradation in the Black Sea has caused serious disruption to the ecosystems, and thus to the provision of goods and services for the peoples of the Black Sea. A brief description of the degradation illustrates these ecological linkages.

The problems of the Black Sea are not so different from elsewhere, but they are more obvious, in part because of the isolated, contained nature of the sea. Of all the seas in Europe, the Black Sea has the longest retention time, 140 years, compared to 80 years for the Mediterranean or one to four years for the North Sea (Stanners and Bourdeau, 1995). The Black Sea also has a large catchment, or drainage area - almost five times the area of the Sea itself (ibid). Thus the problems are generated not only by activities in the six countries on its coastline, but by the actions of the 171 million people in the 17 countries upstream (Zaitsev, 1992).

The problems range from nutrient pollution to introduced species, overfishing, chemical pollution, habitat destruction, loss of biological diversity and more. One of the most serious of these is nutrient pollution, too much nitrogen and phosphorus. Nitrate concentrations increased five-fold between 1960 and 1975 (Mee, 1991, 1992, Stanners and Bourdeau, 1995). This nitrogen comes from a variety of sources: 20% from atmospheric deposition, 30% from agriculture, 20% from industry and 25% from domestic wastewater. More than half of it comes down a single river, the Danube.

These chemical changes trigger biological responses. To an enclosed area like the Black Sea, these vast increases in nutrients spell unmitigated disaster. These nutrients favour and stimulate certain types of microscopic plants or phytoplankton to grow explosively. Many of these phytoplankton are not very good food for the formerly common zooplankton, resulting in changes at higher levels in the food web (Zaitsev, 1992, 1993). The high biomass of phytoplankton also cuts out much of the light reaching the bottom in shallow waters such as the north-west coastal shelf regions of the Black Sea. Some such areas have seen massive losses of shallow water seaweeds as a result. Three species of a red seaweed, *Phyllophora*, have decreased ten-fold on the shelf since the 1970s (Zaitsev, 1992, 1993).

These seaweeds provide critical services and goods to the ecosystem and to people. Like trees in a forest, these plants created a habitat - in this case a unique one inhabited by 118 species of invertebrates such as crabs, shrimp, sponges, worms and snails and 47 species of fishes (Zaitsev, 1993). As the plants disappeared, so too did most of these animals. These *Phyllophora* plants did other things beside create homes. They generated oxygen, as much as $2x10^6$ m^3/da (Zaitsev, 1993). In addition, the seaweeds were collected by people to extract agar (Abbott, 1996). Agar is a useful group of chemicals that are commonly used to thicken food and keep food from separating and to make agar plates for the culturing of bacteria for medicinal and pharmaceutical purposes. In the Black Sea, nutrient pollution triggered a set of changes which led to the demise of the seaweeds, their communities of other plants and animals, their contribution of oxygen and their provision of a useful product for people. Loss of the algae means loss of both goods and services.

Unfortunately this chain of disasters was exacerbated by multiple other simultaneous changes, for example, the accidental introduction of a comb jelly *Mnemiopsis,* via ballast water in the early 1980s. This jelly-fish like creature quickly grew to dominate the seascape: 700 million tons by 1987 (Zaitsev, 1993). It is a voracious predator, feeding on fish eggs, larvae and many species of zooplankton upon which the adult fish depend for their food. Lacking predators of their own, these comb jellies reduced the zooplankton ten-fold, helping to precipitate the crash of economically important fish species including anchovies and sprat (Zaitsev, 1993). The decomposing carcasses of *Mnemiopsis* caused yet another problem: hypoxia, or very low oxygen levels, which in turn triggers yet other problems (Mee, 1991, Zaitsev, 1992, 1993).

These interconnected and synergistic events are but two of the serious problems in the Black Sea. Both the input of nutrients and the introduction of alien species in ballast water are preventable. Moreover, neither problem is limited to the Black Sea. Most coastlines around the world are seeing vast increases in nutrients transported from land-based activities. Introduction of alien species by ballast water is an equally insidious problem around the world and a major contributor to the loss of biological diversity (Carlton and Geller, 1993, Cohen and Carlton, 1998).

The problems of nutrient pollution and alien species are but two of those highlighted in the Black Sea Transboundary Diagnostic Assessment (Global Environment Facility, 1997). This Assessment concludes that the patient - the Black Sea - is indeed very sick. It is not dead, however, but is at a critical point where action is needed. This crisis is an opportunity. The Black Sea Transboundary Diagnostic Assessment and the Black Sea Action Plan (Global Environment Facility 1996) provide a solid basis on which to begin a cure.

The Black Sea Action Plan, signed by the six Black Sea countries on October 31, 1996, is a strong statement of desire to cooperate and address common problems. Formidable challenges remain. There is an urgent need for both individual and institutional ownership of the problem and acceptance of responsibility to ameliorate it - not only by the six countries, and not only by the 171 million people in the catchment, but indeed by the world.

The Black Sea illustrates the need to expand our views of the environment. Its ecosystems have historically provided goods and services to people and other living creatures. Overfishing, chemical pollution, habitat destruction, and loss of biological diversity - in addition to the nutrient pollution and invasive species problems mentioned above - have transformed the Black Sea into an ecological wasteland. This wasteland cannot produce the needed goods and services. Hence:

1. The economy of the Black Sea is beginning to be recognised as an environmental issue: between 1986 and 1993, the fishing industry is reported to have lost 150,000 jobs due to the collapse of fisheries. The tourist industry is reported to have lost $300 million in revenues from the degraded state of the environment (Tass, 1997).

2. Human health in the Black Sea region is being recognised as an environmental issue: toxic viral and bacteriological disease outbreaks throughout the Black Sea region are directly linked to water quality and fish and shellfish contamination (Stanners and Bourdeau, 1995).

3. National security will come to be recognised as an environmental issue in the Black Sea: international tensions rise as resources like fish are in short supply and competition intensifies. The prospect of ecoterrorism is very real. Severe environmental degradation can lead to social disruption and migration.

4. Social justice is an environmental issue in the Black Sea: regional shortages of drugs and medical care to cope with disease outbreaks illustrate the disproportionate impacts of environmental disasters. The consequences of rising prices of food and clean water are borne differentially by poor versus better-off individuals and nations.

Just as people of the Black Sea region depend upon the sea for their well-being and prosperity, so too do people everywhere depend on upon Earth's ecosystems. Understanding these connections is a first step in devising solutions to these problems.

Bibliography

Abbott, I. A. (1996) Ethnobotany of seaweeds: clues to uses of seaweeds, *Hydrobiologia* 326/327:15-20.

Botsford, L. W., Castilla J. C,. Peterson C. H, (1997) The management of fisheries and marine ecosystems, *Science* 277:509-515.

Carlton, J. T., and Geller J. B. (1993) Ecological roulette: the global transport of nonindigenous marine organisms, *Science* 261:78-82.

Chichilnisky, G., Heal, G. (1998) Economic returns from the biosphere, *Nature* 391:629-630.

Cohen, A., Carlton, J. (1998) Accelerating invasion rate in a highly invaded estuary, *Science* 279:555-558.

Daily, G. C. (1995) Restoring value to the world's degraded lands, *Science* 269:350-354.

Daily, G. C. (1997) *Nature's Services*: Societal Dependence on Natural Ecosystems, Island Press, Washington, DC.

Food and Agriculture Organisation (FAO) (1994) Review of the state of world marine fishery resources, FAO Fisheries Technical Paper, 335.

Global Environment Facility (1996) Strategic action plan for the rehabilitation and protection of the Black Sea, Black Sea Environmental Programme, Istanbul.

Global Environment Facility (1997) Black Sea Transboundary Diagnostic Analysis, Black Sea Environmental Programme, United Nations Development Programme, Istanbul.

Hobson, S., Lubchenco, J (1997) *Revelation and the Environment*, World Scientific, Singapore.

Kates, R. W., Turner, B. L., Clark, W. C. (1990) The great transformation, pp 1-17. In:Turner, B. L.,. Clark, W. C,. Kates, R. W, Richards, J. F., Mathews, J. T., Meyer, W. B., eds. *The Earth as Transformed by Human Action*, Cambridge University Press, Cambridge.

Lubchenco, J. (1998) Entering the century of the environment: a new social contract for science, *Science* 279:491-497.

Mee, L. D. (1991) The Black Sea in crisis: A call for concerted international action In: International Workshop on the Black Sea: Focus on the Western Black Sea Shelf, Bulgaria, UNEP/ IOC/IAEA.

Mee, L.D. (1992) The Black Sea in crisis: a need for concerted international action, *Ambio* 21:278-286.

Olson, F. S. (1989) In: Western, D., Pearl, M., eds. *Conservation for the Twenty-first Century*, Oxford University Press, Oxford, p.50.

Postel, S. L., Daily G. C., Ehrlich, P. R. (1996) Human appropriation of renewable fresh water, *Science* 271:785-788.

Schimel, D. S. (1995) CO2 and the carbon cycle, pp 39-71. In: Houghton, J. T., Filho, L. G. M, Bruce, J., Lee, H., Callander, B. A., Haites, E., Harris,N., Maskell K., eds. *Climate Change, 1994: Radiative Forcing of Climate Change*, Cambridge University Press, Cambridge.

Stanners, D. and Bourdeau P. (1995). Europe's environment: The Dobris Assessment. European Environment Agency, Copenhagen.

Steadman, D. (1995) Prehistoric extinction of Pacific island birds: biodiversity meets zooarchaeology, *Science* 267:1123-1131.

Tass News Agency (1997), St. Petersburg, Russia.

Vitousek, P. M., Ehrlich, P. R., Ehrlich, A. H., Matson, P. A. (1986) Human appropriation of the products of photosynthesis, *Bioscience* 36:368-373.

Vitousek, P., Aber, J. D., Howarth, R., Likens, G. (1997) Human alteration of the global nitrogen cycle: Sources and consequences, *Ecological Applications* 7:737-750.

Vitousek, P., Mooney, H. A., Lubchenco, J., Melillo, J. M. (1997) Human domination of earth's ecosystems, *Science* 277:494-499.

Zaitsev, Y. P. (1992) Recent changes in the trophic structure of the Black Sea, Fisheries Oceanography 1:180-189.

Zaitsev, Y. P. (1993) Studies and reviews: Fisheries and environment studies in the Black Sea, General Fisheries Council for the Mediterranean, Food and Agriculture Organisation of the United Nations, 64, Part 2.

Chapter 2

Searching for the Common Ground

Introduction

Laurence David Mee

The three challenging papers in this chapter present the triad of religion, science and the environment from three different perspectives: those of a renowned Christian theologian; a concerned scientist dedicated to communication with the general public; and a Muslim with a profound knowledge of international institutions and their role in contemporary society. The three papers do not attempt to be comprehensive in their analysis but, taken together are highly effective in opening the door to a significant re-evaluation of the linkages between science and ethics today.

Metropolitan John of Pergamon's vision of the partnership which must be forged between science and religion in order to protect the environment, is a powerful and challenging statement. He examines the dichotomy between religion and science, considering science as a child of theology. Some scientists might differ from this perspective, regarding Hellenic philosophy as underpinning the reasoning of both 'mother and child', but certainly religion successfully nurtured early science. Most importantly, the author questions the mechanistic approach of a science which offered the illusion (also to quote Descartes) of 'enjoying, without any effort, the fruits of the earth and all conveniences she may possess'. He extends his criticism to religion which often presents God as ordering humankind to dominate the earth. Metropolitan John examines the way science and theology are moving in convergent paths in their consideration of the natural environment: seeking a holistic approach; searching for a common path between anthropocentrism and ecocentrism; examining the subjectivity of scientific observation and the links between culture and science. He proposes practical ways to continue on this pathway: through fostering the continued development of environmental ethics, perhaps with an international institute of eco-ethics; by encouraging scientists to seek motivation from the understanding of the wider context of their work within our real world; and by developing a new common language on the environment which embraces spiritual and material realities.

David Suzuki presents the vision of a scientist. He reviews his personal enlightenment, from 25 years work as a research geneticist to current interests in the native peoples of the Canadian Pacific Islands. Science, he says, is strong on descriptions but weak on prescriptions because our ignorance is so vast. Traditional and holistic world-views have been debased by our short term and utilitarian thinking and unwillingness to listen to those peers who suggest a harder path for human development. Most environmental scientists will agree with his analysis though may have reservations about the presentation of some of the details. For example, though the level of fragmentary information is an assault, many would argue that the antithesis, a society denied access to information, would be tantamount to a return to obscurantism. Similarly, though global governance (at the service of global economics) puts added pressure on ecosystems, it has diminished the threat of global warfare and offers some hope for solving the awesome environmental problems we have created. Perhaps those who govern are also victims of the deeper problem afflicting our society which is well described by Dr. Suzuki, the lack of love and security in our social and natural environment.

Sadruddin Aga Khan examines personal and institutional responsibility, primarily from the Islamic perspective. He explains how the teachings of Islam exhort stewardship and caring for nature. It warns against disturbing the balance of nature or wasting its produce and creates a strong ethic for sustainability. The reader will surely ponder over why these sound ethic principles have so often been ignored and whether the message has been fully understood. The author describes the 'new values' which have permeated western and eastern cultures, placing emphasis on money and conspicuous consumption. He discusses the impact of globalisation, which diminishes diversity and leads to 'discord and disparity'. The world lacks institutional and legal tools to protect the environment from this new phenomenon though new voices on the international stage are calling for change.

It is interesting that all three papers refer to the issue of authority. Metropolitan John talks of the prestige and respect of science displacing the authority and prestige of religion. David Suzuki, on the other hand, laments that a group of 1,600 scientists, including half of all Nobel prizewinners, was ignored by the media as they warned of irreversible environmental destruction. For Sadruddin Aga Khan, the faiths can speak to their believers with more authority than conservationists. The preoccupation with who has the authority may be misleading. Science is a powerful but imperfect tool used by those who destroy, protect, exploit or nurture, and the values and ethics of scientists stem from the same sources as those of all members of society, though the responsibility of scientists may be greater. Human attitudes and values arise from wisdom and knowledge, often based upon the respect for peers, but supplemented by personal experience. The current arrogance of those who still deny the need for ethics and a change in current life styles is fuelled by our extraordinary but unsustainable efficiency in capitalising nature's resources and the deepening global inequality in human society. For many, authority is in the hands of those who offer a message of plenty, whether they be scientists, religious leaders, salespersons, politicians, stockbrokers or media moguls. The challenge for religion and science is how to convey a message of truth and spirituality into the soul of a society enjoying an irrational pathway towards its own destruction.

Science and the Environment: a Theological Approach

Metropolitan John of Pergamon

Science claims great prestige and respect in our culture. Many people look on it as a power that can cure almost every evil in life. Scientists speak and are listened to with immense authority, because science has managed to claim for itself access to the Truth and the capacity to produce practical results.

Religion, on the contrary, certainly in western societies, has for a long time been deprived of the authority and prestige it enjoyed in the past. It is only in certain circles in our society that a religious truth or doctrine would enjoy more authority than a scientific pronouncement, and this not without a considerable degree of hesitancy on the part of the believer.

What then is the 'use' of religion in tackling such problems as the protection of the environment? Why not leave such matters to the scientists alone? Furthermore, how can science and religion work in common on such matters, given the fact that the former has got factual 'know how', while the latter seems to dwell in celestial places, detached from the realities of life? Is it only by way of complementarity that religion and science can cooperate on environmental issues, the one by contributing inspiration and the other knowledge? Or is there a deeper and organic relationship between these two that can justify their cooperation on such issues?

The dichotomy between religion and science

It can be argued that science, in its modern form, had its origins in religion, particularly in Christian theology. The founders of modern science include such names as Descartes, Leibniz and Newton who were at the same time theologians. In the Jewish tradition, particularly in the later period of the Old Testament when Judaism came into creative contact with Hellenic thought, nature was looked upon by the believer as endowed with rationality. To believe in God would automatically mean to recognise that there is a rationality in creation itself which the human mind could, and perhaps should, explore.

Modern science was born in the thirteenth century when two theological doctrines were crystallised and projected by Christian medieval theologians, namely that creation is rational and that it is at the same time contingent. The belief in rationality led science to the search for causes and explanations that account for the way things are in nature. The faith that creation is contingent, that is, not an extension of God's nature but different from it and possessing its own laws, meant that in order to know nature it is necessary to study nature as it actually is. On the basis of a combination of these two beliefs, modern science was able to embark on the investigation of the way nature is and operates (contingency) with the help of concepts and ideas such as measure, size, quality and cause, borrowed from philosophy (rationality).

Science, therefore, seems to be historically a child of theology. The fact that few of us would be conscious of that today shows that a dramatic change has taken place in the course of history. The child grew up and revolted against its mother and science and religion parted ways.

Tradition represents the Church as the oppressor of scientific research. In condemning Galileo, the archetypal modern scientist, for his Copernicanism, the Church is said to have opened the way to the dichotomy between religion and science. However, the causes of this dichotomy are deeper and more complex and have to do with profound transformations in western thought. I suggest that these transformations are closely related to the question of the environment.

The former Archbishop of York, Dr. John Habgood, theologian and scientist, concludes that the authority and self-sufficiency that science acquired *vis-a-vis* religion was due not so much to a conflict between scientific freedom and ecclesiastic oppression or theological dogmatism, but to the growing spirit of an atomistic and mechanistic approach to nature which has characterised western thought and mentality since the Middle Ages.

Descartes confirms this analysis in words that reveal at the same time the serious effects that the self-sufficiency of science has had with regard to the natural environment. He wrote in his *Discours de la methode*:

> One could find a technique by which the powers and energy, of fire, water, air, the stars, the skies and all the other bodies that surround us, exactly as we know the techniques of our craftsmen, could be applied in the same way to all proper uses so that we might become masters and possessors (maîtres et possesseurs) of nature.

Science has fulfilled Descartes' dream with great success, thanks to its mechanistic approach to nature, its specialisation and its independence from religion, acquiring in this way great prestige and authority in a world seeking happiness. Through its achievement we have indeed become 'masters and possessors of nature'. But is this not precisely the cause of our ecological crisis?

However, religion has its own share in the responsibility. Pushed by science to the realm of mythology and striving to gain recognition and authority in a world seeking happiness, religion in the West entered into competition with science in an attempt to satisfy the growing demands in the market of happiness. It too encouraged human beings to become 'masters and possessors of nature' in the name of either the Bible, which seems to present God as ordering Man to 'dominate the earth', or of a spirituality which exalts human rationality as the divine image in Man over against the debased and sinful materiality of the rest of creation. Religion and science opposed each other in many ways but they seem to have collaborated a great deal, albeit unconsciously, for the destruction of God's creation.

The environment, a challenge to both religion and science

In our own days things are changing rapidly in both science and religion and this is bound to affect the role they will play in the ecological crisis.

First of all, the entire philosophy of specialised knowledge is being questioned in both science and theology, particularly in the former. For example, it is becoming increasingly clear to scientists that zoology and botany are not as clearly distinct disciplines as they were traditionally thought to be: you cannot understand the bee without studying the flowers that determine its life, its whole being and its nature. This environmental interaction and interdependence can be extended *ad infinitum*: everything depends on everything else.

This new holistic approach to knowledge can have important environmental implications for both science and religion. The extinction of a certain species affects the rest of the species. The human being itself is decisively affected by every change in the natural environment. If science moves consistently from the traditional fragmentation of knowledge towards a holistic approach, religion (Christian theology in particular) must revise its views about the human being and admit that humans are inconceivable without their organic relationship with the rest of creation. Christian theology would have to accept the basic claims of the evolutionary ideas of biology, and understand Man as an organic part of the family of animals. There is no essential threat to the Christian faith in accepting the evolutionary theory in its basic principle, that is, the idea that the human being represents the last point in a biological process, although there is no need to accept Darwinism in its detailed description of this evolution. The Bible itself speaks of the creation of Man in the last day and out of natural elements already in existence.

Such an holistic approach would exercise a beneficial influence on people's attitudes to the environment, but this can be effective only if science and religion coincide in their views about the world and the place of the human being in it. The environment can then serve as a catalyst in restoring the organic relationship between science and religion.

Secondly it is becoming clearer to both science and theology that not only is Man dependent on the rest of creation for his existence, but that the reverse is equally true, namely the rest of creation depends on the human being for the realisation and the fulfilment of its existence. Environmentalists need to revise the common assumption that Man needs the rest of creation whereas the rest of creation does not need Man.

Religion - certainly Christian theology - is anthropocentric in its cosmology and would insist that the human being is indispensable to creation. There are signs today that science is moving in the same direction, in the Anthropic Principle which states that the universe is made up in such a way as to make sense only if the human being is presupposed. True humanity requires its organic link with the rest of creation but the latter, too, needs humanity in order to fulfil itself. If the Anthropic Principle is accepted by scientists -and there is evidence that the discussion it has provoked is moving in this direction - a healthy and creative *rapprochement* will take place between religion and science which will have significant implications for ecological thinking.

Thirdly, it is noteworthy that ever since quantum theory won the day in science it has become difficult to eliminate the human being from the process of scientific 'truth'. The observer, we are now told, affects reality in the process of the experiment. What is then left of the traditional subject/object dichotomy? If there is no such thing as a pure 'object' in science, it is no longer possible to operate in scientific research without involving the human person in it. This notion of person, usually regarded as a subject pertaining to sociology, psychology and theology, is now becoming crucial for science too. Science must open its borders to meet with theology and the human sciences if it is to understand correctly its own nature.

Fourthly, it is of crucial importance to note the significance of culture for science. In spite of its tendency to dominate the entire human community, there can be no doubt that western science is influenced by, if not dependent upon, western culture in a decisive way. What would science look like if other cultures were to influence it? Is an African or an Asian science not a conceivable thing? In such a case religion would play a decisive role.

Finally we should note the appearance in our time of hostility towards science because of growing concern for the environment. New Age and all sorts of semi-religious movements are promoting ecological thinking which excludes rationalism and, by implication, science. Where should we stand on this matter? I believe that the environment can be protected in a healthy way only if religion and science open up their boundaries to each other and meet in a creative way. Only by overcoming the traditional dichotomy between these two can we work successfully for the protection of the environment.

Towards a creative cooperation of science with religion

As a theologian I should like to propose the following ways in which one could conceive a cooperation of religion and science for the purpose of protecting the environment:

1. The way of ethics

Religion is recognised as a moral force, having something to say about the way people should behave. This is particularly true of western Christianity. It is becoming customary to seek the view of the Church on how the scientific advances of our time could be controlled so that humanity may be protected from the terrifying consequences of many of the achievements of science. In a period of rapid growth of bioethics, is it not time for the systematic development of a discipline of eco- or environmental ethics, to be given shape and content by theologians and scientists working together? The creation of an international institute of eco-ethics would be a challenge of historic magnitude for the religious and scientific leaders of the world. It would bring together representatives of religious and scientific communities with the responsibility of advising the world's political leaders on environmental ethical questions and to work out a code of environmental ethics which would eventually be invested with the power of international law.

2. The way of motivation

Environmental awareness and behaviour require motivation. Religion can undoubtedly provide motivation for ethical behaviour but science is not necessarily a source of ethical conduct. Surely the promotion of knowledge is not the only motive in scientific work. If it is true to itself, science never believes that its findings are final and its conclusions infallible. The scientist can hardly perform his scientific work properly without realising that the particular knowledge that emerges from it reveals only a tiny part of the world's reality.

The motivation of the scientist and that of the theologian meet at the point of understanding the world as an unbreakable organic unity whose integrity has to be assumed and respected in order for any particular aspect or fragment of its knowledge to be true. Thus, both religion and science, if they wish to be true to themselves, accept that every revelation of reality, whether religious or scientific, can only make sense if the world is respected in its mysterious wholeness.

3. The conceptual way

For centuries scientists and theologians have been using their own, separate, exclusive, esoteric languages. This has made it difficult, if not impossible, for an environmental or ecological language to develop that would involve theologians and scientists at the same time. Technical language may have to remain for internal use on both sides, but when it comes to the environment a common language must be found. For such a common language to develop it will be necessary to work out a conceptual framework shared by both religion and science. We must arrive at a cosmology common to both scientists and theologians and we must also agree on what constitutes knowledge and truth. Agreement will also be necessary on method as well as content. We must stop taking it for granted that theology and religion are about 'spiritual' and 'metaphysical' realities while science is about 'material' matters. The environment is both a spiritual and a material reality.

Conclusion

Science and religion need to forget their traditional opposition towards each other and reach cooperation at the deepest level in order to protect Man's natural environment from the dangers confronting it. Recent developments in both disciplines not only facilitate but demand such a cooperation. There is no doubt that religion and science are, each in its own way, responsible for the ecological crisis we are facing. It is their duty to undo the harm they have done and it is only by working together that they can achieve this.

A Scientist's Approach

David T. Suzuki

I am by training, a geneticist. I graduated from college in 1958, in a period when massive amounts of money were being poured into research and universities. We were taught that science was the most powerful way of knowing. Nothing lay beyond the curious eye of the scientist and, given enough money, we could answer all the questions of the universe. We would acquire the knowledge needed to understand and ultimately control the great forces impinging on us and life would get better and better for all of humankind.

In 1962, I hosted my first television programme about science. I wanted to use this powerful medium to demystify science so that the public would better understand its importance in their lives and they, in turn, would provide better support for research. In that same year Rachel Carson published her seminal book, *Silent Spring* (Houghton Mifflin), in which she issued an important warning about the application of scientific knowledge through technology: every technology, however beneficial, has costs.

We embrace technology enthusiastically because it is deliberately designed to do things for us. When DDT was found to kill insects, the benefits were obvious - it killed pests and chemical companies made money. While geneticists at that time already knew that resistant mutants would soon be selected and render the chemical less effective and ecologists could point out that fewer than one or two out of each thousand insect species are pests to humans so broad spectrum insecticides made little sense, no-one could have warned about the hazards of biomagnification. We only learned about the phenomenon when populations of raptors began to plummet.

Carson's warnings were a potent impetus to the rapid growth of the environmental movement in North America and Europe. Within a decade, a major United Nations conference on the environment was held in Stockholm and the United Nations Environment Programme established. In the following years, names like Bhopal, Exxon Valdez and Chernobyl and new phenomena such as global warming and ozone depletion served to accelerate ecological awareness and concern. Environmentalism expanded in the 1980s from a movement in industrialised countries to a rapidly growing non-governmental community in the developing nations.

Public awareness peaked in 1988 with George Bush's promise to be 'an environmental President' if elected, and Margaret Thatcher's declaration: 'I am a Greenie'. In 1992, the largest number of heads of state in human history gathered in Rio de Janeiro at the Earth Summit to declare that henceforth, humanity would take a different path. From that point on, 'sustainable development' would be the goal while all areas of human activity would measure and incorporate the ecological costs. As if to punctuate the importance of the

Earth Summit, in November, 1992, the Union of Concerned Scientists released a document entitled *World Scientists' Warning to Humanity*. Signed by 1,600 senior scientists from 71 countries and including more than half of all living Nobel prizewinners, this remarkable document began:

> Human beings and the natural world are on a collision course. Human activities inflict harsh and often irreversible damage on the environment and on critical resources. If not checked, many of our current practices put at serious risk the future that we wish for human society and the plant and animal kingdoms, and may so alter the living world that it will be unable to sustain life in the manner that we know. Fundamental changes are urgent if we are to avoid the collision our present course will bring about.

It went on to describe the crisis in population, atmosphere, oceans, water, soil, forests and species loss, then grew stark:

> No more than one or a few decades remain before the chance to avert the threats we now confront will be lost and the prospects for humanity immeasurably diminished. We the undersigned, senior members of the scientific community, hereby warn all humanity of what lies ahead. A great change in our stewardship of the earth and the life on it is required, if vast human misery is to be avoided and our global home on this planet is not to be irretrievably mutilated.

This report went virtually ignored by the world's news media. A suggestion by more than half of all Nobel laureates that we may have as little as ten years to avoid a global catastrophe was judged not newsworthy. Today, the media have pronounced the environment a dead issue and their attention is focused on the global economy. What has happened?

Tim Flannery, author of *The Future Eaters*, suggests that the current destructive spasm is not a new phenomenon. The great inventiveness and enormous capacity for memory that distinguished us from other animals gave us the ability to use our brains to project into the future. It also enabled us to overexploit our surroundings as, long ago, we migrated into new habitats. As 'future eaters', Flannery suggests we are a species that diminishes the options for future generations by a tendency to overexploit. What brings us into balance is the accumulation of knowledge about an area over countless generations. That knowledge was contained in the world-view of a group, the sum total of thinking, observing, inventing and imagining. And the insights acquired were profound.

As Claude Levi-Strauss says:

> I see no reason why mankind should have waited until recent times to produce minds of the calibre of a Plato or Einstein. Already over two or three hundred thousand years ago, there were probably men of similar capacity (*Man the Hunter,* Aldine, 1968).

In a world-view, everything is interlinked - the stars, clouds, rivers, forests, fish, human beings. The past, present and future are all part of a continuum. Nothing exists independently or in isolation. Each world-view is inseparably tied to a specific place and time and is profoundly local. Human beings were deeply embedded in this world, charged with enormous responsibility to keep it all going. Thus, we had to say the right prayers, observe the appropriate rituals and ceremonies and behave properly, if the earth was to continue to provide.

Today traditional world-views have been shattered to such an extent that we no longer recognise the ecological ramifications of what we do: driving a car two blocks instead of walking, expecting fresh fruits and vegetables year round or purchasing any consumer item. This disconnection has been accelerated in this century by a number of factors which continue to atomise our universe while often blinding us to the consequences.

Population

The explosive increase in human numbers, from 2 billion to more than 5.6 billion in my lifetime, is made up of a majority of people who were born after 1950. For them, the unprecedented and unsustainable changes that have occurred all their lives seem normal and necessary and must be continued.

In 1900, more than 95% of all human beings lived in rural, village communities. There were only 14 cities with a million people or more, none larger than 7 million. Today there are 400 cities of a million or more and the ten largest have over 11 million people. We have become an urban species and have lost the connections that inform us that Mother Earth pays a price to absorb our sewage and garbage and deliver us water, energy and food.

Information

The vaunted information superhighway is, in fact, a superHYPEway. We are assaulted with information fragmented into snippets devoid of context, history or time.

Government

Governments now govern on an unprecedented scale, charged with making decisions that reverberate across the entire planet. Politicians pay attention to a constituency of voters and financial supporters, but children do not vote, nor do generations yet to come, and the species-chauvinism of our perspective disenfranchises fish, trees, air and oceans. Politicians do not speak on behalf of non-human resources, only on behalf of the people who want to exploit them.

Economics

Economics, a human construct devised to serve people and their societies is today reified and the relationship inverted. The economy externalises those things that keep us alive - the atmosphere, underground aquifers, topsoil, ozone layer, and because economists have such faith in our capacity to discover, replace or substitute resources, endless growth is deemed not only possible but necessary. We are using up the basic biological capital on which all life flourishes and which is the rightful legacy of all future generations.

Science and technology

Finally we come to the role played by science and technology. Scientists focus on parts of nature. We bring a fragment into the laboratory, separate it from all else, control the factors impinging on it and measure everything within it. In so doing, we acquire a profound understanding of that isolated piece of nature yet often lose a sense of the context that made it interesting or important in the first place. Ever since Newton's time, we have perceived the universe as a giant clockwork mechanism whose elementary components can be examined to derive a comprehensive understanding of the whole and eventually be pieced back together like a giant jigsaw puzzle

Modern physics has demonstrated that this does not work. Pieces of nature interact synergistically when combined so that new properties emerge from the complex that cannot be anticipated from the properties of the separate parts. In response it has been suggested that even if we cannot predict behaviour of higher levels of organisation from the properties of underlying components, we may find principles that provide us with predictive capabilities at each level of organisation. As a geneticist, I have been astounded and exhilarated at the enormous insights and manipulative dexterity acquired in my field, but we are often so intoxicated with our discoveries that we confuse an incremental insight with being the key to opening the door to understanding and control of everything around us. We create highly simplified models of reality so that 'experts' can claim to have enough knowledge to manage forests, nuclear disposal, fish, topsoil, air and water. Yet each insight gained only serves to emphasise the enormity of our ignorance

What, for example, is a forest? Some foresters confuse a monoculture of planted trees with a forest when in fact, a forest is the totality of air, water, soil and a vast community of organisms, most of whose components are not even identified. We understand next to nothing about how species are interlocked to make a vibrant, productive and resilient community of living things. Estimates of the number of species on Earth range from a few million to a hundred million. A widely accepted figure of ten million is acknowledged to have been arrived at very conservatively. Harvard's E.O. Wilson indicates to date that about 1.4 million species have been taxonomically identified (*Naturalist*, Simon and Schuster, 1997). That merely means that someone in a laboratory has given a name to a dead specimen. It does not mean that we know anything of substance about that species' numbers, geographic distribution, habitat, food, behaviour, reproduction or interaction with other species.

This is not to denigrate the astounding progress and discoveries that have been made by scientists. I simply suggest that we temper our enthusiasm with some humility. I graduated as a fully licensed geneticist in 1961 and I thought then that I knew everything there was to know about genetics and would set the world on fire. For 25 years, research in genetics was my obsession and greatest joy. Today, when I tell students the ideas we had about chromosome structure and genetic regulation in 1961, they gasp or laugh in disbelief. Today, most of the best ideas of 1961 can be seen for what they are - wrong, irrelevant or just unimportant. Indeed, that is the way science progresses, by proving that most current ideas are incorrect or wide of the mark.

The great strength of science is in *description*. Scientists make discoveries everywhere because we have so much to learn. But where science is weak is in *prescription* of solutions because our ignorance is so vast. At best, we can manage *ourselves*, learning to respect the boundaries of nature and pulling back in order to allow nature to replenish the things that support us.

My period of working as a full time bench scientist remains the happiest time of my life, full of moments of joy, discovery and sharing. I am proud of what I have done and grateful for having had the opportunity to work as a scientist. However, it is clear to me that we have so severed the connections that link us with everything that we forget there is a bottom line far more profound than matters economic. Now in a new area of grassroots environmental activism, I have discovered a new way of seeing the world through the eyes of the Haida, the aboriginal people of Haida Gwaii, the islands Canadians call the Queen Charlottes. For these people, and for every group of aboriginal people I have encountered, their identity is inseparably connected with the land. The land embodies their history, culture, purpose and meaning.

From the Haida I have begun to understand a fundamentally different way of looking at the ecocrisis. We are animals and as biological beings, we have an absolute need for earth, air, fire and water - they are the sacred elements of the ancients and ought to remain so today. Without them, we cannot live. And what ensures the abundance and purity of these primary elements is the collective diversity of living things on the planet.

We are also social animals and we have a deep need for love. Without it, we are psychically and physically crippled. The best way to ensure conditions that encourage the love that humanises us is to provide conditions that allow stable families and communities - full employment, justice and security. Numerous studies endorse this.

Finally, we are profoundly spiritual beings with a need to know our relatives, the other life forms that share the planet with us. Even though we will all die, we are comforted by the knowledge that nature will persist and flourish. We need to have sacred places that ecologists call wilderness sanctuaries. We have to know there are cosmic forces beyond our understanding or control.

If these constitute our real bottom line, the foundation on which communities and human activity rests, then our priorities must change.

Personal and Institutional Responsibility

Sadruddin Aga Khan

Personal responsibility and the Islamic perspective

In a world characterised by interdependence and all encompassing global 'civilisation', as well as by diversity and unique individuality, shared ethical principles governing the interactions of human beings and their world are essential, if the well-being of the larger community of life, upon which we all depend, is to be protected.

One of the great moral and spiritual issues of our day is the massive and increasing destruction of Creation. At the same time we are witnessing a spiritual (in the widest sense of the word) re-birth and need to re-discover communion with nature. The growth of the Green movement is but one manifestation of this trend, yet the philosophy of sustainable use lacks a moral and ethical underpinning that would give it credibility, integrity and consistency.

Only if there is a widely diffused ethical concern, which individuals deliberately make part of their own consciousness and seek to transmit to their children, is there a chance of re-creation of the balance between Humankind and nature. Religion, far more than governmental bodies or pressure groups, has a fundamental role to play in this respect. The Faiths can speak to their believers with more authority than conservation alone and while the implementation of environmental management depends on the existence of appropriate legislation, legislation becomes more effective, enforceable and useful when it emanates from a nation's creed - when it represents its cultural and intellectual heritage.

All the great religions of the world - as well as the spirituality of first peoples - have maintained a tradition of conservation, respect for nature and the environment, and kindness to other living creatures. These tenets of 'environmental theology' are little known and seldom discussed, much less widely observed or practised. They are, however, perhaps more relevant today than ever. Ancient wisdom, Taoism, Zen Buddhism, Master Eckhart, St. Francis of Assisi, Ibn al A'rãbi and mystics of all Faiths speak of the family of Man as a link in the chain of all beings on this Earth. Indeed, mystics always tend to relate to each other in a spirit of true ecumenism, due to their common bond with Creation. And Creation is nature.

The essence of Islamic teaching is that the entire Universe is God's creation. Not only does Islam mean submission to the will of Allah; the root of the Arabic word also means peace. Each individual must find peace with Allah, with himself, with humanity and all Creation. Islam promotes the integrity of the Earth. Unity cannot be had by discord, by setting one need against another or letting one end predominate over another; it is maintained by balance and harmony. Therefore Muslims say that Islam is the middle path and we will be answerable for how we have walked this path.

The injunctions of the Qur'an and the Prophetic traditions about the balanced order of nature were expounded in an ethic of ecology in *Ikhwan al-Safa* in its famous epistle on animals and Man. Several chapters of the Qur'an are named after animals, and many hadiths of the Prophet commend kindness to them. Mohammed is reported to have said, 'all creatures are God's dependants and the most beloved to God, among them, is he who does good to God's dependants.'

The inherent relationship between faith and the environment finds symbolic expression in the various regular rituals. Prayer is timed according to the Earth's planetary movement; Hajj, or pilgrimage, is timed according to the seasons and water is an essential element used to prepare the believer both physically and spiritually.

Indeed, water - the focus of our symposium - is recognised by Islam as the basis of all life on our planet. Even the word Shariah - the Islamic law which guides human interactions - means the road to the watering place. Water is the source of life and the Shariah is the source of knowledge about life, death and the Hereafter.

For the Muslim, Humankind's role on Earth is that of a *khalifa*, vice-regent or trustee of God. We are God's stewards and agents on Earth. In the words of the Prophet, 'the world is green and beautiful and God has appointed you his stewards over it.' The *khalifa* is answerable for his or her actions, for the way in which he or she uses or abuses the trust of God. This concept of planetary trust and stewardship is recognised by Judaïsm and Christianity as well as by many other religions. It is also explicit and implicit in major instruments of international law and has been adopted extensively by secular thinkers as a metaphor to describe the role of Humankind.

For the Muslim, each generation of Humankind is ethic-bound to leave behind a wholesome, sustainable environment. 'He it is who has made you inheritors of the Earth' says the Qur'an. 'It is He who produces gardens and crops diverse. Eat of their fruits... pay the due thereof... and be not wasteful.' Nor does the status of vice-regency give ground for arrogance or tyranny: 'And walk not in the earth with insolence; certainly thou wilt never tear the earth open.'

Humans are but the managers and the administrators and the Qur'an warns that *fasad*, or corruptions, will be prevalent on Earth as a result of human disobedience of God. Muslims believe that all creation follows one law, the Law of Allah. If this law is not obeyed, the balance of creation is disturbed and according to Qur'an interpreters, disobedience will lead to:

> Tribulations, catastrophes both on land and sea... dryness of the land, many fires, many drowned and a reduction in the blessings of God...lower crops in both food, plants and fruits.

These sobering and sadly topical words may be instructively contrasted with the Biblical warning:

> The Earth mourns and withers, the world languishes and withers; the heavens languish together with the Earth. The Earth lies polluted under its inhabitants; for they have transgressed the laws, violated the statutes, broken the everlasting covenant' (Isaiah 24:4-5).

This closely mirrors the foreboding of the Apocalypse of St John. The Christian scriptures also state that Mankind's fate depends on protecting and preserving the Earth's life-support systems:

> For that which befalleth the son of men befalleth beasts, even one thing befalleth them: as the one dieth, so dieth the other, yea, they have all one breath; so that a man hath no pre-eminence above a beast' (Ecclesiastes 3:19-21).

Islam teaches us that we have been created by God and that we will return to God for judgement: that we are accountable for our deeds as well as our omissions. The *khalifa* will render an account of how he treated the trust of God on the Day of Reckoning. The notion that describes the accountability of the *khalifa* is *akhirah.* Islam is the guidance of how to live today so that we can face the *akhirah*; it is the message which informs us of what will be involved in that reckoning. The Qur'an states specifically that the responsibility of vice-regency will be checked by God himself to see how it has been performed.

Unity, trusteeship and accountability (*tawheed, khalifa and akhirah*), the three central pillars of Islam, are also the pillars of the environmental ethics of Islam. These notions must not be reduced to matters of personal piety: they must guide all aspects of life.

The most positive potential impact of the Shariah may, indeed, be in the field of the environment where it has a uniquely strong and comprehensive tradition on which modern Muslim legislators might build. Let us not forget that Muslims had formulated a bill of legal rights of animals as early as the thirteenth century and in Turkey Muslims had established environmental and cultural associations and initiated work on environmental issues long before the Ministry of Environment was established.

Islam should be a complete way of life, an ethical system providing the bearings for all our actions. We must integrate true Islamic values into our very being. We must judge our actions by them. They furnish us with a topical world-view which enables us to ask environmentally appropriate questions, draw up the right balance sheet of possibilities, properly weigh the environmental costs and benefits of what we want and what we can do within the ethical boundaries established by God, without violating the rights of His other creations.

As the late Professor Al-Hafiz Masri noted:

> It is a thousand pities that, while Islam has a wealth of guidance to offer, our Islamic countries are going on blindfold repeating the mistakes that the West had initiated after the Industrial Revolution.

It is my heartfelt hope and belief that Muslims will yet create a true Islamic alternative, a caring and practical way of being, doing and knowing. The philosophical and spiritual bases do exist, but more guidance needs to be available to the faithful and those who govern them. There is, for example, a relative paucity of practical information about conservation in Muslim countries compared to the proliferation of such materials available in the West.

Institutional responsibility

In spite of the warnings and sound intentions broadcast by the Club of Rome, the Brundtland Commission, the Earth Summit and a host of similar initiatives, the institutional response has consistently proved frustratingly inadequate to meet the challenges of our times.

The current fascination - almost to religious fervour - with globalisation will not replace development aid, yet the balance of power seems to be shifting dangerously from the hands of elected representatives towards mega-corporations and a new breed of anonymous globocrats.

With precious little accountability or public scrutiny, decisions are being made in the name of progress and liberalisation which carry highly negative side-effects for human health, quality of life, the environment and our collective conscience. The new values place the emphasis firmly on money and conspicuous consumption. Governments and non-governmental organisations tend to follow the corporations in applying management criteria and profits and taxes have replaced the common good.

Globalisation is the opposite of bio-diversity. It brings in its wake the dangers of discord and disparity - the erosion of the middle path so crucial, as we have seen, to the Islamic concepts of unity, harmony and the integrity of the Earth. True happiness is based on spiritual as well as material values and there is clearly a need to counterbalance economic interests with parallel emphasis on the social, environmental and ethical perspectives. Why could there not be a codification of environmental conventions - or even a new World Environment Authority vested with the same level of prestige, efficiency, authority, credibility, means and influence as the World Trade Organisation? Why is it that Environment Ministers are currently all too frequently viewed as poor cousins to their Trade counterparts?

I have, however, been greatly encouraged by the new, more forceful language which is creeping into documents like the report of the Committee on the Environment, Public Health and Consumer Protection of the European Parliament prepared for the recent United Nations General Assembly Special Session to review Agenda 21. This underscores the reciprocity between economic and social development and protection of the environment which is vital to any truly viable concept of sustainable development.

More importantly, the report recognises the need for the European Union to set an example, notably in terms of financial commitment, and highlights the importance of clearly defining priorities, setting precise targets and creating specific tools for achieving measurable results, thus moving beyond rhetoric towards an operational phase which is already tragically overdue. It also proposes an International Environment Court to give teeth to environmental legislation.

In the Black Sea region, as elsewhere, most of the problems, and the analyses needed to design remedies for them, are known. The tragedy is that this knowledge is not adequately applied - either because there is no economic interest or capability, or because concerned individuals feel powerless to intervene. There is an outstanding legacy of marine research which must be revitalised. There is a wide range of institutions that could be actively involved, including schools, churches and businesses. There is also the need for specific leadership from inter-governmental and non-governmental organisations. For all the many merits of the Black Sea Strategic Action Plan, success in the Black Sea regions will ultimately depend on individuals, singly and collectively. The greatest potential rests with human will and in this respect the message of Islam is wholly pro-active. As the prophet says, 'On Doomsday, if anyone has a palm shoot in hand, he should plant it'.

Chapter 3

Living Resources

Introduction

Graeme Kelleher

Despite the implications of the title of this chapter, 'living resources', the papers here do not treat the ecosystems in which people live as entities that are separate from them and available for indiscriminate use by our species. Rather, this chapter, as do others, recognises the physical and spiritual inseparability of humanity and nature.

Richard Chartres and Paul Erhlich deal with humanity in nature on a global basis. They show how, spiritually and materially, modern society has become disconnected from the intimate relationship with nature which lay at the root of earlier civilisations and which still pervades the thinking and attitudes of some less developed communities, particularly some of those called indigenous communities. The long term consequences of the modern attitude, both for humanity and nature, are described; already clearly visible throughout the world, many of them are quantified here.

Yuvenaly Zaitsev, Ulrich Niermann and Boris Alexandrov focus on the Black Sea. They summarise the present condition of the Sea and the creatures that live in it and show how human activities have contributed, and continue to contribute, to the deterioration in its ecological quality. Possible solutions to the many evident problems are identified.

It is clear from all that is written here, and in other chapters in this book, that a profound change in human attitudes and commitment to nature is vital if the Black Sea and the human communities that depend upon it are to survive and prosper. It is also clear that this change is not going to occur with sufficient commitment from the people in the 17 countries that affect the Black Sea unless the change is of the degree of profundity that normally flows from religious conviction, rather than from the intellectual conviction that arises from science. The same conclusion applies to the global environment and the world's human population.

This chapter illustrates the need for religion to inspire people and for actions to be based on scientific knowledge and methods so that the living resources of the biosphere can be protected for the immediate and distant future.

Sacrifice and Covenant, Prometheus and Noah

Richard J.C. Chartres

The waters of the Black Sea are heavily freighted with memories of some of the stories which come from the springtime of human culture. The narrative of the Great Flood exists in various forms but perhaps the best known is recorded in the Book of Genesis and has Noah as its central character. Noah and his family together with representatives of 'all flesh' were preserved from the inundation by the Ark which eventually, we are told, came to rest on Mount Ararat to the east of the Black Sea.

When it was safe to do so, the survivors emerged. God established a covenant with Noah and his sons that he would never again bring destruction on the earth. In Noah, God blessed not just some chosen race but the whole of humanity, and he made a covenant which embraced 'all' and not just human flesh. In particular he set a limit to what human beings could eat. They were forbidden to consume the blood of the beasts and fowls. The ancients believed that life itself was in the blood and so the divine prohibition established a balance between human appetite and the integrity of the created order, which was not to be simply for consumption.

When I mentioned the story of Noah to a scientist friend, his brow clouded. He began to brief me about certain disputes about whether the veritable ark had been identified in the region of Ararat. Here is clearly an initial problem about paradigms. I have no difficulty in believing that there were great floods in pre-history but I am frankly not very interested in the Noah story if it is little more than the record of a meteorological catastrophe. The fundamentalist approach, which insists that, had a camera been available, it would have been possible to make a home video of the Ark's progress exactly as it is described in Genesis, is a relatively modern and unhappy tribute to the dominance of the scientific way of establishing and describing truth by reference to evidence accessible to a detached observer and ultimately quantifiable. While the achievements of this way of exploring our environment are not in doubt, in the case of the Noah story and other spiritual writings the method leads us into a dead end. It sometimes increases our information but does not add to our *wisdom.*

The climax of the Noah story also illustrates another profound paradigm shift. God promises never again to bring destruction upon the earth and appoints the rainbow as a sign of his promise. Clearly the world was experienced as reverberating with the divine footsteps and any threat was seen as coming from the divine wrath. For most people, even religious believers, in western countries, this way of looking at the world has yielded to one in which the main source of anxiety comes from the use human beings make of their atom-splitting power. The sign which hangs over our generation is not that of the rainbow but of the mushroom cloud of 1945.

The divine was being evacuated from the western world-view long before this century. Before the seventeenth century western culture shared the almost universal view of nature which persists still in many parts of the world, that nature was sacred and that the divine signature was visible. However, from the seventeenth century onwards nature was seen as an inanimate machine to be dominated and exploited by man. Nature came to be soulless and it is instructive that the strategy of denying soul has been applied in other cases where man has wished to dominate and possess, in particular in the debates about the souls of slaves and native people in the early modern period. Descartes frankly avowed that the motive behind his method of reasoning was to make man 'maître et possesseur de la terre'.

What happened next is anticipated in another ancient story associated with the Caucasus mountains to the east of the Black Sea. The story of Prometheus fascinated both Renaissance and Romantic thinkers in the West. When the gods and human beings came to be divided it was through the agency of Prometheus, whose name means Fore-thinker. He cut up an ox for sacrifice, concealing the choicest portions beneath the paunch of the beast for himself and his people. He covered the ox bones with an alluring veneer of gleaming fat and then called out: 'Zeus, most famed and greatest of the eternal gods, choose thou the portion that thou desirest'. Some say that Zeus saw through the imposture but the fraudulent sacrifice and the trickery of Prometheus made a divide between gods and men.

There are also various versions of the story of Prometheus' theft of fire from the gods. As a punishment, Zeus had Prometheus chained to the highest peak of the Caucasus. An ancient vase painting shows him with a pillar driven through his middle being attacked by an eagle. The bird may represent the mental world of the Fore-thinker and it is significant that every day the eagle comes to devour his liver, the seat of the intuitive life. All that is consumed by day however, grows back again by night. It was prophesied that Prometheus would be unbound in the thirteenth generation.

The sixteenth century French philosopher Charles de Bouvelles compared the emancipated man with Prometheus, whose seizure of power, which allowed human beings to change their nature, was presented in a positive light. In his book *de Sapiente* written in 1509 he pictures man as 'no longer part of the universe but as its eye and mirror; and indeed as a mirror that does not receive the images of things from outside but that rather forms and shapes them in itself.'

With the de-animation of nature, however, and the substitution of dominance for connectedness in the relation between the Fore-thinker and the universe around him, the ambitions of Promethean man become literally inordinate. At the same time the development of the rational calculating faculty at the expense of other ways of knowing has led to distress and disorder in the liver of the intuitive life. The capacity to be aware of the sacred and animated character of the cosmos has been impaired. Although the word sacred is still in use, it tends to be used in a sentimental way. How can objects of our own thought be sacred in the same way that once the world permeated by the divine was held to be sacred?

These cultural developments go some way to explaining why, with a few exceptions, western theology has tended to concentrate on the development of environmental ethics rather than attempting to re-assert the religious view of nature. There is, however, a renewable energy which flows from the recovered awareness of the sacredness of nature, which is not available when nature is regarded as a mental projection of Promethean man.

I want to suggest that a revival of this way of living in nature, knowing oneself to be a pontifical person, a bridge builder and a part of the natural order, offering unforced sacrifices and observing the covenant which prohibits inordinate appetite, sheds some light on a puzzle which is often unremarked because it is so common.

There have been many meetings of scientific and religious bodies which have issued solemn declarations built on a rigorous scientific appraisal of the ecological challenges facing humanity as we near the end of this millennium. Some can still be studied with profit. As long ago as 1971 my own Church solemnly declared that pollution was not just a serious mistake but a 'blasphemy'. The first European Christian Assembly held in Basel in 1989, which brought together leaders of all the main churches, chose 'engendering a sense of responsibility to the environment' as one of its main themes. The Assembly made some very specific recommendations in line with the Brundtland report about the need for industrialised countries to reduce energy consumption and 95% of the delegates voted for this programme. However, very little has been done subsequently in the life of the churches themselves to achieve the targets. We know what should be done but we do not do it.

As well as investigating the ecological problems of the Black Sea region with scientific rigour, I believe that we must also focus on the problem of translating what we know into what we do. As the psychologist C.G. Jung observed, 'Mere appeals to ethical fraternity can never evoke in human beings the slightest trace of that age-old power which drives the migrating bird across the sea'.

Working away at codes of ethics is not to be despised and there is some important work going on at present to develop a 'global ethic'. We need to establish internationally agreed standards and of course secular ethicists are partners in the task. Preaching about personal ethics, however, is only part of the necessary response to the ecological challenge of our day. Spiritual practice is needed of the kind which enlivens awareness and connects with inexhaustible spiritual energy. Some people are finding their way into the reality and energy disclosed by this awareness but there is a great intellectual challenge confronting western theology in particular to remove the obstacles and inhibitions which hamper and deflect the spiritual progress of so many modern people.

The importance of education has already been stated. However, alongside the analytical way of seeing the world, an adequate education should also develop the faculty for contemplation which reveals other beings in a new luminosity. 'Sin' was defined by one of the saints of this century, Simone Weil, as 'Inattention'. We need an education in contemplative attention.

Theology and science have to negotiate a new partnership in the problems facing us all in our common ark. We ought not to divide the one reality and seek, as Moltmann the distinguished German theologian has expressed it, 'peaceful co-existence at the price of mutual irrelevance'.

We need to look again at the philosophical foundations of the way we perceive the world and to reflect on the question posed by the poet T.S. Eliot in one of the choruses from 'The Rock':

> Where is the wisdom we have lost in knowledge? Where is the knowledge we have lost in information?

There is also an institutional as well as an intellectual challenge. The institutional problem goes some way to explaining the disappointing response to so many well-formulated statements from conferences and commentators in the past. Scientists have successfully created an international community and to a large extent a common language but the faith communities sometimes seem less disposed for dialogue with one another than they were in previous generations. The task is to translate the moral imagination and awareness of individuals into institutional forms which can have an impact on the emerging global forum. Earth summits and international scientific congresses need a more adequate partner to express the contribution of the faith communities. As the Brazilian proverb puts it, 'If I am the only person who dreams, it remains only a dream but if several of us live the same dream it can become a reality'.

The point about living the dream is the most important of all. I have sketched an intellectual and an institutional response but we also need the awareness which transforms behaviour with spiritual practice. A profound response to the ecological challenge facing us will be difficult without the recovery of the capacity to sacrifice on the part of people in the richer part of the world. Awareness of what we are doing with the divine gift of life in this world can bring home the injustice that is being done to the world's poor and, at the same time, the injustice of expecting our children to pay for the style of life we enjoy now. Sacrifice has always meant giving up some cherished possession and referring it to the divine, so that it can be seen in a different light and received back as a gift, charged with a new valency and capable of sustaining us in our role as bridge builders, pontifical persons in the creation.

Again, however, we need not so much exhortation as a new breed of spiritual entrepreneurs, rather like the earliest friars who embraced the insecurity and fear of the mediaeval cities in a vulnerable but joyful lifestyle. These eco-friars need to be able through their spiritual practice to organise and knit together the frayed strands of community. Such people will need, like Mother Theresa, to express themselves in visible symbols rather than in tracts. Certainly in the western Christian world they will need to re-invigorate the Sabbath festival as a celebration of sustainability and 'enoughness', rather than present it as a pause before the next bout of frantic activity or as a dowdy day of gloom and restriction. The Sabbath offers both relief from the pressures of the passing moment and an anticipation of the crown of creation, a foretaste of the divinely intended state of sustainable equilibrium. The living resources which have always flowed from the Sabbath in the Jewish tradition need to be available to refresh us all.

A Holistic Approach to the Role of Human Beings in Ecosystems

Paul R. Ehrlich

Human exceptionalism is a plague in our society. The idea that *Homo sapiens* is somehow separate from nature rather than an integral part of it is widespread. Many people believe that food comes from the supermarket, water pours from the tap, climate is controlled by air conditioning and heating, and space is conquered by automobiles and space shuttles. According to this view, human beings have dominion over the planet; they can use nature in any way they please with no risk to themselves. Human beings are not, after all, *animals*.

Human beings as animals

This common view is certainly wrong. A fine fossil record and detailed comparisons of human anatomy, behaviour and genes with those of our living relatives tell the story. We need only trace back our family tree for about 30 million years (less than half-way back to the age of the dinosaurs) to find the common ancestor we share with old world monkeys. For the past decade scientists have known that evolutionarily we are positioned right in the middle of the great apes. About five million years ago our ancestors split off from the evolutionary line leading to the chimpanzees, while gorilla ancestors left the chimp-human line about seven million years ago (Horai et al., 1995). Genealogically we are more closely related to chimpanzees than chimpanzees are to gorillas.

Our physical resemblance to the other great apes is apparent to any unbiased observer. Our genetic resemblance is spectacular: over 98% of the genetic material of human beings and chimpanzees is identical. Our fundamental perceptual abilities are very chimp-like (Matsuzawa, 1985b, Matsuzawa, 1985a, Matsuzawa, 1990) and our behavioural resemblance is often spectacular as well.

Among the other apes (and elsewhere in the animal kingdom) can be found behaviours that once were thought to be characteristic only of human beings. Chimpanzees use a wide variety of tools and weapons, manufacturing some of them. Like people, they raid neighbours' territories and sometimes deliberately hunt down and kill members of rival 'tribes' - a primitive form of warfare (Wrangham and Peterson, 1996). Chimps also have culture (Boesch, 1996), a body of non-genetic information that is passed from individual to individual and from generation to generation, as exemplified by observational learning of tool use (Matsuzawa, 1996) and in the development of local 'tool kits,' which show each community to have a different cultural tradition (McGrew, 1992). Furthermore, the 'handedness' of human beings, once thought to be a unique characteristic of *Homo sapiens*, is also found in apes (McGrew and Marchant, 1996). Other 'human' behaviours are also characteristic of the societies of our close relatives. Chimpanzees live in patriarchal societies, as do people. Male gorillas often kill offspring of their mates that they had not fathered, a behaviour unhappily not uncommon in human societies. And, like males of most other mammals, men often try to inseminate more than one mate.

How are people different?

This is not to say that people are just the same as other apes. There are, of course, some striking morphological differences. Human beings are the only apes with a fully upright posture. We are also relatively hairless, have brains roughly three times as large as those of the other great apes, and female human beings are different from all other of our relatives in having large breasts prior to bearing children.

Perhaps the biggest difference in kind between human beings and other apes is our use of speech and language. There is little evidence that our close relatives use vocal communication significantly more than other mammals. If they do communicate extensively with each other, it is not through sounds (Mitani, 1996). There are other big differences in kind between human beings and other apes. We are the only animals that use fire, clearly show empathy and have religion. Another major difference in kind is the human use of contraception. We are the only organisms that deliberately choose to limit our reproduction at levels that do not maximise an individual's genetic contribution to future generations (Ehrlich et al., 1995).

There are also some astonishing differences of degree. No other organism remotely approaches the cultural endowment of our species, indeed of any normal human individual. Each of us has access, through spoken language and imitation and in most cases through writing and electronic media, to a vast store of knowledge. Furthermore, our culture clearly evolves, resulting in the development of behaviours that no isolated individual could invent. Such cultural evolution may occur in chimpanzees but it is unknown in other mammals (Boyd and Richerson, 1996). Related to this is the enormous elaboration of technology by human beings.

The role of ecosystems in human life

Human beings, like all other organisms, exchange materials with their environments. Thus they are elements in natural ecosystems. Ecosystems consist of interacting combinations of organisms and their physical environments. People breathe the oxygen that was added to Earth's atmosphere by the photosynthetic activities of multitudes of plants and microorganisms, past and present – and all of the energy that powers human metabolic activities is derived from food that comes directly or indirectly from photosynthesis. Human beings drink water circulated and purified by a hydrological cycle that involves the activities of myriads of plants that both transfer water from Earth to atmosphere and anchor soils that mediate water flows. They live in climates stabilised by those ecosystems, a stability that is also essential to the success of agriculture. In short, humanity is utterly dependent on goods and services it receives from natural ecosystems (Daily, 1997).

Human roles in ecosystems

However, the participation of *Homo sapiens* in ecosystems is so different in degree from that of other organisms that it amounts to a difference in kind. Because of its evolving store of culture, and especially the development of agriculture some 10,000 years ago, humanity has become the dominant animal species on the planet (Ehrlich and Ehrlich, 1990). At the time of the agricultural revolution there were about five million people scattered over Earth – fewer than now reside in the San Francisco Bay area. By the time of Christ, that number had expanded to some 200 to 300 million, about the present population of the United States. There were a billion people by the early nineteenth century, and about two billion when I was born in 1932. Today we have reached almost six billion - if I live two more years, the human population will have tripled in my lifetime. The rise of our species to total dominance is the most dramatic event on our planet since its collision with a comet some 65 million years ago. That disaster caused the extermination of the dinosaurs along with many other organisms, and (ironically) opened the door for the rise of the mammals and us.

The human economy, the 'metabolism' of that dominant animal, has altered Earth in ways that will be evident for hundreds of thousands - or even millions - of years. This will not be the result of our building and construction achievements, most of which will not persist for even 10,000 years; rather, lasting impacts will follow from the massive extinction of species that accompanies human activities. We are also significantly altering the gaseous quality of the atmosphere, interfering, for example, in major ways with the carbon and nitrogen cycles (Vitousek et al., 1986). Humanity has now approximately doubled the annual natural rate of addition of nitrogen to the terrestrial nitrogen cycle, creating potentially serious problems for the maintenance of soil fertility, accelerating losses of biodiversity, contributing to acid deposition and enhancing the greenhouse effect (Vitousek et al., 1997a).

The most serious potential consequence of interfering with major nutrient cycles may be rapid climate change, of which there are already ominous (if non-definitive) signs such as the unusual scale and persistence of the El Niño phenomenon and an increasing frequency of extreme weather events in general (Karl et al., 1995). It is important to note that the paleoclimatic record does not suggest gradual climate change in response to factors such as changing atmospheric composition, but rather sudden shifts from one quasi-stable state to another (Schneider, 1995), such as the 'Younger Dryas', an episode about 12,000 years ago when the post-glacial warming was interrupted for a 500-year cold snap. It has long been clear that humanity has been running a gigantic and dangerous experiment on itself (Ehrlich, 1968). If we are lucky, negative feedbacks may damp out much of the impact of global warming, and change may be gradual or almost non-existent. If we are unlucky, changes may be sudden and catastrophic, especially for agricultural production, which is already showing signs of failing to keep up with the population's growth.

Some gradual changes have already been long under way that many ecologists find even more threatening than the possibility of rapid climate change. Human beings have already degraded over 40% of the land surface, causing it to lose about 10% of its potential direct instrumental value - its ability to yield benefits such as agricultural, forestry, industrial, and medicinal products (Daily, 1995). A conservative estimate of the amount of soil that is lost to wind and water erosion annually is 25 billion tons (Brown and Wolfe, 1984). People have appropriated over half of the accessible flows of fresh water (Postel et al., 1996) and around the world are pumping aquifers at rates much faster than those at which they are recharged (Ehrlich et al., 1995). Human beings have already destroyed about half of all tropical forests, and the current rate of depletion is probably about 1.5 to 2.5% per year. In the United States, virtually all natural grasslands and about half of all wetlands have been destroyed since the time of Columbus, and about three-quarters of all the coastal mangrove wetlands of Thailand, Pakistan, and India have disappeared, mostly since the Second World War (Ehrlich, 1995). People now mobilise trace metals such as zinc, lead, and chromium at many times the rate of natural processes such as erosion and volcanic outgassing, and spill much more petroleum into the oceans than do natural seeps (Ehrlich et al., 1977).

Most crucially, humanity now uses, co-opts, or destroys some 40% of terrestrial net primary productivity (NPP). NPP is the sum of all the photosynthetic efforts of plants, minus what the plants need to support their own activities. One can think of it as the potential food supply of all animals and other non-photosynthesising organisms Vitousek et al., 1986). This diversion of NPP to humanity's own ends, along with anthropogenic habitat destruction and the transport of invasive plants and animals, is the cause of what may prove to be *Homo sapiens'* most serious impact on Earth's ecosystems - the depletion and destruction of populations and species of other organisms. Humanity has launched a vast extinction episode that may well exceed that which ended the age of the dinosaurs, 65 million years ago. The result may be virtually permanent impairment of the ability of ecosystems to continue to supply civilisation with absolutely essential natural services, such as amelioration of the climate, provision of fresh water, flood control, soil creation and maintenance, waste disposal, nutrient recycling, crop pollination, and pest control.

Rates of extinction of species now appear to be 100 to 1,000 times those that prevailed for most of the millions of years until humanity's rise to dominance (Vitousek et al.,

1997b), and populations may be disappearing at 5,000 or more times prehistoric rates Hughes et al., 1997). Recovery of biodiversity from the late Cretaceous disaster took many millions of years.

In short, humanity is rapidly depleting the 'natural capital' upon which all of us depend: most critically, deep, rich agricultural soils, groundwater accumulated during the last ice age, and biodiversity. The latter loss is likely to prove most serious in the long run - especially if, as now appears likely, a high level of biodiversity is crucial to the delivery of ecosystem services, particularly those that support agriculture. With the added possibility of rapid climate change, the outlook for civilisation, from an ecological viewpoint, could be considered decidedly grim.

The socio-political environment

People are not only embedded in ecosystems, they are also embedded in societies. Today societies are subjected to two countervailing trends. On the one hand, in economic and cultural globalisation, the rich world, increasingly segregated from the poor, is being united by global corporations that answer to no government (Korten, 1995). At the same time, the disintegration of the Soviet empire and the general failure of communism has contributed greatly to a violent ethnic fragmentation that is especially serious in the poor world (Barber, 1995). It is far from clear where these trends are leading us, but it is certain that they will have a substantial impact on how we deal with the ecosystems that support us. It is obvious, for example, that if fragmentation becomes the overriding trend, cooperation to solve global problems will be extremely difficult to achieve. On the other hand, the increasing power of transnational corporations could easily accelerate the growth of the metabolism of the dominant animal and intensify the unsustainable pressures already being applied to humanity's life-support systems.

The solution

It may not be too late for humanity to avert a vast ecological disaster and successfully transition into a sustainable society, but the task will not be simple. Population growth must be halted and a slow decline begun to an 'optimum' population size. Fortunately birthrates are falling now in many parts of the world. Wasteful consumption in rich nations must be reduced in order to allow for needed growth in poor nations. The reduction of consumption while increasing the quality of life is technologically feasible; but technological feasibility is not enough. Our socio-political systems need to undergo dramatic revision in the direction of increasing equity at all levels if sustainability is to be achieved. One of the overriding reasons is that the cooperation needed to solve global environmental problems is unlikely to be achieved in a world divided into 'haves' and 'have nots.' The challenge is immense, and we must move rapidly to meet it.

Bibliography

Barber, B.R. (1995) *Jihad vs. McWorld*, Ballantine Books, New York.

Boesch, C. (1996) The emergence of cultures among wild chimpanzees, pp. 251-268 In. Runciman, W.G., Smith, J.M., Dunbar, R.I.M., eds *Evolution of Social Behaviour Patterns in Primates and Man*, Oxford University Press, Oxford, UK.

Boyd, R., Richerson, P.J. (1996) Why culture is common, but cultural evolution is rare. In Runciman, W.G., Smith, J.M., Dunbar, R.I.M, eds *Evolution of Social Behaviour Patterns in Primates and Man*, Oxford University Press, Oxford, UK.

Brown, L., Wolfe, E (1984) *Soil Erosion: Quiet Crisis in the World Economy*, Worldwatch Institute, Washington, DC.

Daily, G.C. (1995) Restoring value to the world's degraded lands, *Science* 269: pp.350-354.

Daily, G.C., ed (1997) *Nature's Services*, Island Press, Washington, DC.

Ehrlich, P.R. (1968) *The Population Bomb*, Ballantine Books, New York.

Ehrlich, P.R. (1995) The scale of the human enterprise and biodiversity loss. In Lawton, J.H., May, R.M. eds *Extinction Rates*, Oxford University Press, Oxford.

Ehrlich, P.R., Ehrlich, A.H. (1990) *The Population Explosion*, Simon and Schuster, New York.

Ehrlich, P.R., Ehrlich, A.H., Daily, G.C. (1995) *The Stork and the Plow: The Equity Answer to the Human Dilemma*, Putnam, New York.

Horai, S., Hayaska, K., Kondo, R., Tsugane, K., Takahata, N (1995) Recent African origin of modern humans revealed by complete sequences of hominoid mitochondrial DNAs, *Proceedings of the National Academy of Sciences* 92: 532-536.

Hughes, J.B., Daily, G.C., Ehrlich, P.R. (1997) Population diversity: its extent and extinction, *Science* .

Karl, T., Knight, R., Plummer, N. (1995) Trends in high frequency climate variability in the twentieth century, *Nature* 377: 217-220.

Korten, D.C. (1995) *When Corporations Rule the World*, Kumarian Press and Berrett-Koehler Publishers, West Hartford, CN and San Francisco, CA.

Matsuzawa, T. (1985a) Color naming and classification in a chimpanzee, *Journal of Human Evolution* 14: 283-291.

Matsuzawa, T. (1985b) The use of numbers by a chimpanzee, *Nature* 315: 57-59.

Matsuzawa, T. (1990) Form perception and visual acuity in a chimpanzee, *Folia Primatologica* 55: 24-32.

McGrew, W. (1992) *Chimpanzee Material Culture: Implications for Human Evolution*, Cambridge University Press, Cambridge.

McGrew, W.C., Marchant, L.F. (1996) On which side of the apes? Ethological study of laterality of hand use. In McGrew, W.C.,. Marchant, L.F, Nishida, T., eds. *Great Ape Societies*, Cambridge University Press, Cambridge, UK.

Mitani, J.C. (1996) Comparative studies of African ape vocal behaviour. In McGrew, W.C., Marchant, L.F., Nishida, T., eds. *Great Ape Societies*, pp. 241-254. Cambridge University Press, Cambridge.

Postel, S.L., Daily, G.C., Ehrlich, P.R. (1996) Human appropriation of renewable fresh water, *Science* 271: 785-788.

Schneider, S.H. (1995) The future of climate: potential for interaction and surprises. In T.E. Downing, ed. *Climate Change and World Food Security*, pp. 77-113. Springer-Verlag, Heidelberg.

Vitousek, P.M., Aber, J.D., Howard, R.W., Likens, G.E., Matson, P.A., Schindler, D.W., Schlesinger, W.H., Tilman, D.G. (1997a) Human alteration of the global nitrogen cycle: sources and consequences, *Ecological Applications* 7: 737-750.

Vitousek, P.M., Ehrlich, P.R., Ehrlich, A.H., Matson, P.A. (1986) Human appropriation of the products of photosynthesis, *BioScience* 36: 368-373.

Vitousek, P.M., Mooney, H.A., Lubchenco J., Melillo, J.M. (1997b) Human domination of Earth's ecosystems, *Science* 277: 494-499.

Wrangham, R., Peterson, D. (1996) *Demonic Males: Apes and the Origins of Human Violence*, Houghton Mifflin, New York.

The Black Sea: Status and Challenges

Yuvenaly P. Zaitsev

From the ancient Greek age to the 1950s the marine environment of the Black Sea was more or less stable. However, studies have shown that since 1950 to 1960 there have been significant transformations in the Black Sea ecosystem. These changes have put an end to a millennial period of relatively stable ecological balance and have marked the beginning of a new era in Black Sea biology and ecology. Its main manifestations are eutrophication (over fertilisation) chemical pollution, accidental introduction of exotic species, bacterial pollution, coastline changes and overexploitation of marine living resources.

Man-made eutrophication of coastal waters and shelf zones has been the most harmful impact on the sea both in terms of scale and its consequences (Zaitsev 1992a, 1993). The major sources of nutrients entering the sea include fertilisers from fields, discharges from animal husbandry, untreated urban sewage and atmospheric fall-out. The area most affected by eutrophication is the north-western part of the sea, where the three largest rivers, the Danube, Dniester and Dnieper are incoming.

Increase in nutrients in the river discharge triggered the mass development of phytoplankon and an expansion in the area of its blooms. This phenomenon has been thoroughly studied by Ukrainian, Romanian and Bulgarian scientists. A large algal biomass provides good feeding conditions for plant-eating animal plankton (zooplankton) and a sharp increase in the numbers of some zooplankton species occurred in the 1970s and 1980s. The most obvious reaction of zooplankton in eutrophic conditions was the mass development of so called gelatinous organisms, whose bodies consist of up to 98 to 99% water.

A high number of the large jelly-fish, *Rhizostoma pulmo,* was observed in the coastal zone of the north-west Black Sea in the late 1960s and early 1970s. Later, in 1973-1974, the population of this species dropped back and almost immediately there was an outburst in the population of another jelly-fish, the moon jelly, *Aurelia aurita.* Until the 1960s the total average biomass of *Aurelia* in the Black Sea was 0.7 million tons. In the late 1970s it rose to an average of 220 million tons and in the early 1980s it reached 500 million tons. It is difficult to say what the fate of *Aurelia* would have been and how its relations with other species would have developed had not another gelatinous species, the exotic ctenophore *Mnemiopsis leidyi*, unexpectedly appeared in the early 1980s. By the late 1980s it had reached a total biomass of close to one billion tons. The *Aurelia* population collapsed almost immediately, the biomass of other zooplankton dropped sharply and the catches of the main Black Sea commercial fish, anchovy, sharply decreased.

The precise role that *Mnemiopsis* played in these events and whether it was the only culprit is still to be resolved, but it is commonly held responsible for much of what happened (Zaitsev and Mamaev, 1997).

One of the consequences of eutrophication is the decreasing in transparency of water as the result of greater quantities of phytoplankton, *Noctiluca*, jelly-fish, ctenophores and organic debris. The amount of sunlight reaching the bottom in the shelf zone is directly dependent on the transparency of the water. The result has been severe degradation of bottom communities of algae.

The fate of the Zernov's *Phyllophora* field, considered the largest aggregation of red agar-bearing algae of *Phyllophora* genus in the world ocean, is significant in this sense. In the 1950s, this meadow of red algae was occupying an area of 11,000 km^2 in the central part of the north-west shelf. The total biomass of algae was variously estimated at 7 to 10 million tons. By the early 1980s the algal field had shrink to 3,000 km^2 and its biomass to 1.4 million tons. By the early 1990s, the field was only 500km^2 and its biomass did not exceed 300,000 to 500,000 tons (Zaitsev, 1992 (a)) This decline represented not only the loss of valuable raw material which was harvested for the extraction of agaroid, a gelling agent, but also the disappearance of an important generator of oxygen at depths of 20 to 60 metres where Zernov's *Phyllophora* field was situated. It has been estimated that the photosynthesis in this field in its unthreatened state produced up to two million cubic meters of oxygen per day (Zaitsev,1992 (a)). Moreover, *Phyllophora* was the nucleus of a bottom community known as the *Phyllophora* biocoenosis which included up to 120 species of invertebrates and 50 species of fish. Many of these animals have evolved a protective dark red and reddish coloration and this has led to the concept of '*Phyllophora* fauna'. With the disappearance of key species, the entire biocoenosis collapsed.

Similar changes have also taken place in the coastal zone. The large brown alga *Cystoseira barbata*, which is intolerant of eutrophication, has virtually disappeared. It is no longer found along the coast of Ukraine (except at Crimea) or Romania and has become scarce along the Bulgarian coast. *Cystoseira* is the keystone species of the *Cystoseira* biocoenosis, which included dozens of species of invertebrates and several species of fish.

The most damaging effects of man-made eutrophication for bottom animals (zoobenthos) was the appearance of zones with reduced (hypoxia) or no (anoxia) dissolved oxygen in the near-bottom layer of water.

Large-scale mortality of bottom-living animals was first recorded in the Black Sea in August-September 1973 in the north-west shelf across an area of 3,500km^2. Subsequently mass mortalities became an annual occurrence and the area of the north-west shelf over which they extended increased to 30,000 to 40,000 km^2 (Zaitsev, 1992b). It was established that between 100 and 200 tons of bottom organisms died from a depletion of oxygen per square kilometre of shelf. In the period between 1973 and 1990 the loss of biomass amounted to 60 million tons of bottom-living animals including 5 million tons of fish, both young and adult, commercial and non-commercial species.

The death of vast quantities of benthic animals is not just a question of the disappearance of important commercial fish species, with the resulting economic losses. The environmental damage is also enormous and includes the consequences of the mass mortalities of filter-feeding organisms such as the mussel.

As to chemical pollution of the Black Sea, this kind of man-made influence is not a basin-scale phenomenon, but in some coastal areas (near so-called industrial hot spots) there are elevated concentrations of heavy metals. Some higher concentrations of the insecticide lindane were found in the Danube discharge zone but most samples were comparable with those in the Mediterranean.

For oil pollution, the Danube river accounts for 48% of the 110,000 tons of oil entering the Black Sea each year. Most of the remainder is introduced from land-based sources through inadequate waste treatment and poor handling of oil products. Concentrations of this pollutant in the surface layer of the western Black Sea is one order of magnitude higher than in the Mediterranean.

Bacterial pollution of coastal marine waters is connected with untreated sewage. In the Black Sea coastal region approximately 10.4 million people discharge an estimated 571 million m^3/year of sewage into the Black Sea. As a result, there are regular beach closures in some Black Sea hot spots and increasingly frequent outbreaks of serious water-borne diseases such as cholera and hepatitis A. Not all Black Sea bathing waters are below standard, however.

The accidental introduction of harmful exotic species is a very specific kind of biological pollution. Because of its unpredictability, the involuntary introduction of species is one of the most serious man-made impacts. It can have considerable biological, ecological, economic and social repercussions.

The Black Sea has become home for a large number of alien marine organisms. Three of the most significant examples are *Rapuna thomasiana* from the Sea of Japan, the soft-shelled clam, *Mya arenaria,* from the Atlantic Ocean and the exotic ctenophore, *Mnemiopsis leidyi.*

Rapuna thomasiana, a large gastropod mollusc and notorious predator that feeds on oysters, mussels and other bivalve molluscs, was recorded in Novorossiysk Bay in 1946. In the 1950s it depleted the Gudauta oyster bank on the Caucasian shelf and began to feed on the mussels on the Crimean shelf and in Bulgarian waters. In the 1970s it penetrated the Sea of Marmara. At first the only factor limiting the *Rapuna* population growth was the local souvenir industry. In the 1980s it was discovered that there is a demand for *Rapuna* meat on the Asian market. Massive commercial catches of the *Rapuna* along the Turkish and Bulgarian coasts were undertaken and after several years the number of this snail diminished to such a level that its commercial fishing became unprofitable.

Mya arenaria, a large bivalve mollusc, was recorded in 1966 on the Odessa Gulf. It became widespread on the Black Sea shelf, especially at low salinities and out-competed the local small bivalve *Lentidium mediterraneum*, which had been an important food source for the fry of many species of fish. Today, *Lentidium*'s habitat is restricted to a narrow coastal strip. In the 1970s, *Mya* biomass became greatly diminished because of hypoxia.

The 1980s was the golden age of *Mnemiopsis leidyi*. Now it continues to be a dominant species in the pelagic zone but its population seems to be in decline, at least in the north-western Black Sea, where commercial catches of anchovy were seen to be gradually increasing over 1995 and 1996. The population of the anchovy-eating bird shearwater, *Puffinus puffinus*, has also increased since 1994.

At the edge of the sea, different man-made changes occur. This coastline zone is very sensitive to human impact. It has been suggested that as a narrow peripheral strip whose edge represents only an infinitesimal part of the total surface of the Black Sea, the coastal zone may be ignored when considering measures to protect the sea's living resources. But such a view is mistaken for two fundamental reasons. Firstly, the narrow coastal zone is not peripheral but a focal point containing the highest densities and the greatest variety of organisms and is home to biological processes of critical importance. Secondly, the coastal zone is the part of the marine environment where people come into direct contact with the sea and where the health of people is directly dependent on the health of the sea.

During the last 20 to 30 years the Black Sea coastal zone has been affected by a number of direct and indirect anthropogenic influences which have produced different responses from the ecosystem.

Large scale anti-landslide and coast protection works have been undertaken in the Odessa area and as a result destructive landslides have been halted, the beach area increased ten-fold and a recreational infrastructure has been developed. However, these measures have created stagnant zones in the sea where the sand became silted and the bottom biocoenoses fundamentally changed.

The construction of dams for hydropower stations on several Caucasian rivers has led to a sharp reduction in solid discharges and an acceleration of coastal erosion processes.

Unmanaged exploitation of the Black Sea biological resources is another kind of man-made impact on its ecosystem. This happened to the largest flatfish, the turbot, *Psetta maeotica*, whose commercial fishing was prohibited in Russia and Ukraine in the 1980s. Overfishing was also the main reason for the sharp decline in sturgeon (*Acipenseridae*) populations and four species of fishes have been included in the Ukrainian Red Data Book (1994) of endangered species.

In the 1950s the population of all three species of Black Sea cetaceans was estimated at one million individuals. By the middle of the 1960s their populations had dropped to 300,000 individuals as a result of an increase in catches. In 1966 an agreement was reached by the former USSR, Romania and Bulgaria to stop dolphin fishing and in 1983 it was stopped in Turkey. The present population of these mammals needs to be estimated, but according to visual observations it continues to decline. The Black Sea population of the monk seal, *Monachus monachus,* is about to become extinct. The major factor critical to its fate is not overcatching (this species is internationally protected) but the lack of adequate sites for reproduction on shore. This is a very careful and apprehensive animal which avoids populated and reconstructed coastal zones.

While overcoming these objective and subjective factors is a constituent part of the national and regional implementation of the Strategic Action Plan for the Rehabilitation and Protection of the Black Sea, it is a vast spiritual problem as well as a practical one. Success depends to a considerable extent on the understanding, readiness and persuasion of the general public. For natural and especially for human reasons, the modern Black Sea has become a kind of sombre symbol of impacted marine areas in the world.

Bibliography

Alexandrov, B.G., Zaitsev, Y. P. (1996) Ukrainian National Report on Black Sea Biodiversity (submitted).

Komakhidze, A., Mazmanidi, N. (1966) Georgian National Report on Black Sea Biodiversity (submitted).

Konsulov, A., Konsulova, T. (1996) Georgian National Report on Black Sea Biodiversity (submitted).

Ozturk, B., (1996) Turkish National Report on Black Sea Biodiversity (submitted).

Petranu, A., (1996) Romanian National Report on Black Sea Biodiversity (submitted).

Strategic Action Plan for the Rehabilitation and Protection of the Black Sea. Istanbul (1996).

Volovik, S.P. (1996) Russian National Report on Black Sea Biodiversity (submitted).

Zaitsev, Y. P. (1992a) Recent changes in the trophic structure of the Black Sea, Fish Oceanogr., 1, (2).

Zaitsev, Y. P. (1992b) Ecological state of the Black Sea shelf zone, Ukrainian coast (a review), *Gidrobiolog, zhurnal* 28 (4).

Zaitsev, Y. P. (1993) Impacts of eutrophication on the Black Sea fauna, General Fisheries Council for the Mediterranean, Studies and Reviews, Rome, FAO, 64.

Zaitsev, Y., Mamaev, V. (1997) *Marine Biological Diversity in the Black Sea, A Study of Change and Decline*, New York, United Nations Publications.

Black Sea Fisheries: Restoring, Protecting and Sharing

Ulrich Niermann

With a dramatic decline in landings from 850,000 tons in the mid 1980s to a low of 250,000 tons in 1991 and a serious decline in valuable fish species, the collapse of the Black Sea fishery from 1989 onwards has been severe. The situation was worsened by the existing economic and political difficulties in the region and although there was some recovery after 1992 with a total landing in 1995 of 517,000 tons, the main improvement has been to the low value anchovy stock while higher valued species have remained depressed or continued to decline.

The total amount of landings at first point of sale (based on 1994 prices) fell from US$ 626 million in 1985 to around US$ 250 million in 1993-95 and whereas the Turkish fleet has been increasing in size, the other fleets have been declining. The bulk of the catch was taken by Turkey (358,018 tons), while the other countries caught only 54,500 tons (Bulgaria: 5,500 tons; Romania: 3,000 tons; Ukraine, Russia, Georgia together: 46,000 tons).

A total of about 168 fish species occur in the Black Sea and from 1960 to 1979 about 26 commercial species were caught at a rate of 10,000 tons per year. By the beginning of the 1990s only five species were left (anchovy, sprat, horse mackerel, whiting, and bonito) and anchovy has been the most important fish in terms of fish catch since the mid-1980s.

It is difficult to relate the collapse of the fishery to just one impact and we must consider a range of different causes, man-made and natural, which have influenced the fish catch. There have been changes to the environment with damming, dredging and dumping, with chemical pollution, eutrophication, overfishing, the introduction of new species and changes caused by climatic variability.

The damming of rivers blocks the migration of fishes such as sturgeon to their spawning places in the upper rivers (Zaitsev, 1992). Thus the number of offspring declined strongly, which led to a decreasing sturgeon stock in the Black Sea. Dredging and dumping in the sea, especially in coastal areas, changes the structure of the sediment, which leads to a change in the bottom flora and fauna and finally to a decline of former valuable fish species.

Chemical pollution, pollution with heavy metals and oil pollution in particular has a tremendously negative effect on the development of fish eggs and larvae. The most polluted area is the north-western shelf of the Black Sea; this is also the most important spawning and feeding area for fishes like anchovy, turbot, atlantic bonito and the blue fish, primarily because of river-induced productivity, and because the underlying shallow and broad shelf area, inhabited by algae, provide an excellent environment for bottom fishes (Zaitsev, 1992).

Since industrialisation in the 1960s, the northern shelf area has been subject to dramatic changes primarily as a result of increased nutrient input via the major rivers, the Danube, Dneiper, and Dniester, flowing into its northern region. The increase led to a higher unhealthy primary production and to dramatic changes in the species composition of the phyto- and zooplankton. Species with a high food quality for fishes were replaced by small opportunistic species with a low food value. Jelly organisms, like the planktonic algae *Noctiluca scintillans* and the jelly-fish *Aurelia aurita* appeared in high densities. During the 1970s and 1980s there was a significant drop in the large pelagic and demersal fish species like mackerel, bonito, blue fish and turbot (Zaitsev, 1992).

The spawning activity of the fishes decreased strongly at the north-western shelf. Many anchovy eggs were observed in the open sea, but after 1992 the bulk of eggs was found in the southern and south-eastern area of the Black Sea. It could be assumed that the improved food conditions in the open sea due to eutrophication had resulted in a shift of the spawning grounds of the anchovy to the open sea. Increasing eutrophication of the former mesotrophic coastal areas of the southern Black Sea may be the reason for the expansion of the spawning grounds from the Bosphorus area to the eastern Black Sea coast of Turkey. Similarly the anchovy stock of the Sea of Azov shifted to the south, because the conditions for spawning became worse due to desalination, the result of hydrological manipulation on the in-flowing rivers, like Kuban (Niermann et al, 1994; Gücü, 1997).

These changes in the environment and planktonic communities of the Black Sea could be assumed to be the basic conditions for the outbreak of the recently accidentally introduced new jelly-fish species the ctenophore *Mnemiopsis leidyi*, which originated from eutrophic lagoons in North America. Reaching an estimated biomass of a remarkable 800 million tons wet weight in the summer of 1989 this voracious ctenophore consumed a considerable fraction of the zooplankton, which had been the food for pelagic fish and their larvae previous to the appearance of *Mnemiopsis* (Vinogradov et al., 1989, 1992; Mutlu et al., 1994). Investigations by Schulman (1997) have shown that the condition of the anchovies was very poor and that most individuals were starving by the end of the 1980s. The sharp decline in anchovy eggs and larvae from 1988 to 1991, reflecting the collapse of anchovy catches for the whole of the Black Sea in 1989, coincided with the mass development of the new invader *Mnemiopsis*. *Mnemiopsis* showed the typical pattern of a new coloniser: after mass development in the years 1989 and 1990 with 2 kg m^{-2} in the open sea and average numbers of 4.7 kg m^{-2} at the north-western shelf, the numbers and biomass dropped to a moderate level by 1991. Up to 1996 the biomass of

Mnemiopsis fluctuated between 0.2 and 0.4 kg m^{-2}. It is very striking that the anchovy stock recovered after 1992, when the biomass of *Mnemiopsis* declined to moderate levels. But on the other hand, during the mass development of *Aurelia aurita* during the 1970s, reaching about the same biomass in terms of carbon as the *Mnemiopsis* bloom, the impact to the anchovy stock was very little. This suggests that there were other causes for the decline of the anchovy.

Overfishing could be another reason for the decline of the Black Sea fish stock. At present there is no international management of the fish stocks of the Black Sea. As an example let us consider again the anchovy. During winter the anchovy migrates towards the warmer waters of the Turkish coast and forms very dense schools in a narrow coastal band. The anchovy schools are extremely vulnerable to the Turkish fishing vessels, which had become very well equipped with sonar since the middle of the 1980s, able to detect, surround, and catch the fish schools. As a result, the size of the spawning stock was so reduced that not enough young fish were produced to ensure the future stock.

Beside these man-made impacts, the effect of physical and meteorological changes at the end of the 1980s seems to have had an effect as well. Striking changes in the air temperature, sea surface temperature, salinity, river discharges and sea level were obvious from the mid-1980s to the beginning of the 1990s in the western Black Sea. After a cold period from 1984 to 1988 the air and sea temperature increased from 1988 to 1993 with exceptionally warm years from 1989 to 1991. The river discharges, especially that of the River Danube, decreased drastically by the middle of the 1980s, increased again until 1988 and decreased again during 1989/90. The sea level displayed similar patterns with a strong decline in the middle of the 1980s and a fast increase from 1988 to 1990. The salinity was well below the 40 year average (1955-1995) from 1987 to 1992 (Polonsky et al., 1997).

These meteorological and hydrological changes were obvious in different regions of the world in the same period. The comparison of anchovy catches in different upwelling systems of the world showed that the anchovy stock of South Africa, Benguela and California also collapsed in the same period as the anchovy stock of the Black Sea, yet without the impact of *Mnemiopsis* (Niermann et al., in prep).

Anchovy is a fish species which feeds at the lower end of the food chain, on zooplankton species. Therefore it is strongly dependent on the fluctuation of its prey species. If these prey species decline, it could be assumed that the anchovy stock should decline as well. Studies of other regions of the world demonstrated that the biomass showed similar fluctuation patterns as in the Black Sea: high biomass in the 1970s, a decline of the biomass in the beginning of the 1980s, a rise in biomass during the mid-1980s and a drastic decline at the end of the 1980s and beginning of 1990s. After 1993 there was an increasing trend. Similar trends in zooplankton were found in the Baltic Sea, in the North Sea, in the North East Atlantic, in the waters off California and in fresh water lakes such as Lake Windermere in Great Britain and the Bodensee in Germany. It is possible that this negative trend in zooplankton biomass led to the decline of the anchovy stock in different regions of the world. The biomass of the commercially exploited benthic and pelagic species also decreased by the end of the 1980s (Niermann et al., in prep).

The same fluctuation patterns in different environments could imply that a certain climatic signal occurred at the end of the 1980s and the beginning of the 1990s. Indeed, by the end of the 1980s, many hydrographical and meteorological weather patterns changed over the northern hemisphere.

- In the open Black Sea the air and the sea surface temperature decreased after 1986, were extremely low in 1987, raised in 1988 until 1991 and decreased again until 1994. Changes in salinity were obvious in the western Black Sea by the end of the 1980s. Similar changes in the hydrographic parameters were observed in the Baltic Sea, North Sea and other regions of the northern hemisphere.
- Changes in hydrographical and meteorological parameters, air temperature, atmospheric pressure, precipitation and changes in the strength of westerly winds, were recorded by the end of the 1980s in the North Sea and the English Channel.
- Distinct changes were measured in the NAO (North Atlantic Oscillation; the mean sea level pressure difference between Azores and Iceland), SOI (Southern Oscillation Index; the mean sea level pressure difference between Darwin and Tahiti), ENSO (Southern Oscillation (El Niño) Index), and ALPI (Aleutian Low Pressure Index) at the end of the 1980s.
- The shift of the northern wall of the Gulf Stream was far southwards during 1986-1988 compared to the previous and subsequent years.
- The oscillation of the global air temperature, which has a period of 65 to 70 years had its lowest amplitude during the mid-1980s.
- The sun activity, measured in terms of sun spot numbers, which has a period of 11 to 12 years, was very low during the mid-1980s and raised back to high values at the end of the 1980s.
- The moon's orbit crossed the ecliptic in 1988 (lunar tide, with a period of 18.6 years), which causes an extreme peak in tidal forces and has well documented effects on the sedimentation rates in the Wadden Sea.

Although the short and long term atmospheric variability in the Eastern Mediterranean and Black Sea regions is well known, hypotheses on teleconnections with global atmospheric events have only recently been put forward. It has become obvious that the climate over the North Atlantic influences the Black Sea hydrology, which indicates a global ocean-atmosphere coupling. As changes in the zooplankton community in the late 1980s were evident in all the seas under consideration, a climatic impact could have triggered the changes in the zooplankton community in the Black Sea as well, causing the conditions for the outburst of *Mnemiopsis leidyi* and the decline of the anchovy stock.

That the catch of many Black Sea fishes has recovered since 1993 could be due to the following reasons:

1. When the catch of one fish species declines, it becomes uneconomical to exploit this fish stock further. The fishermen switch to other fish species in other seas;
2. Due to the bad economic situation, many factories in the north-western Black Sea region had to close, which led to a reduction of pollutants and fertiliser (nutrients) released into the sea and which may have improved the environment, especially in the nursery and feeding grounds;
3. The climatic regime which occurred during the period at the end of the 1980s and beginning of the 1990s came to an end and the production of fodder zooplankton improved again until the mid 1990s. In addition, the number of the new invader, *Mnemiopsis,* decreased to moderate levels after its outburst in 1988-1990; this is typical for new invader species.

In order to manage and to restore the fish stocks of the Black Sea it is essential that we assess the most important impacts on the fish. In conclusion, I would recommend that:

- An international fish stock management should be established to regulate the fish catches;
- An international monitoring system should be established to observe the natural fluctuation in the ecosystem and to control the results of man-made changes such as damming, dumping, dredging, chemical pollution and inputs via rivers and air;
- Man-made impacts to the ecosystem (chemical pollution, oil pollution, fertiliser, herbicides, changing of the environment) should be reduced;
- The production of chemicals which affect the atmosphere and the climate should be avoided.

This last topic in particular should be taken extremely seriously, as research on climatic changes and on teleconnections with global atmospheric events is at its very beginning. It is known that many species of the phytoplankton and zooplankton, the basic food of the fishes, are very sensitive to small changes of the climate and if the composition of the fodder zooplankton changes, than it is bound to have an impact on the anchovy and other fish stocks of the Black Sea.

Bibliography

Gücü, A., (1997) Role of fishing in the Black Sea ecosystem. In: Ozsoy & Mikaelean (eds.): Sensitivity to change: Black Sea, Baltic Sea and North Sea, NATO ASI Series, 2. Environment Vol.27. Kluver Academic Publishers (pp: 516): 149-162.

MacLennan, D.N., Yasuda, T., Mee, L. (1997) Analyses of the Black Sea fishery fleet and landings, BSEP-PCU, Dolmabahce Sarayi, II Harekat Kosku, 80680 Besiktas, Istanbul.

Mutlu E., Bingel, F., Gücü, A.C., Melnikov, V, Niermann, U., Ostr, N.A., Zaika, V.E. (1994) Distribution of the new invader *Mnemiopsis sp.* and the resident *Aurelia aurita* and *Pleurobrachia pileus* populations in the Black Sea in the years 1991-1993, ICES Journal of Marine Sciences 51, 407-421.

Niermann U., Kideys, A., Kovalev, A.V., Melnikov V. (1977). In prep. Long-term fluctuation of the zooplankton of the open Black Sea in comparison to other regions of the world, NATO-ARW: Environmental Degradation of the Black Sea: Challenges and Remedies, Constanta, Romania, October 6-10.

Niermann U., Bingel, F., Gorban, A., Gordina, A.D., Gücü, A.C., Kideys, A.E., Konsulov, A., Radu, G., Subbotin, A.A., Zaika, V.E. (1994) Distribution of anchovy eggs and larvae (*Engraulis encrasicolus* Cuv.) in the Black Sea in 1991 and 1992 in comparison to former surveys, ICES Journal of Marine Sciences 51, 395-406.

Niermann U., Greve, W. (1997) Distribution and fluctuation of dominant zooplankton species in the southern Black Sea in comparison to the North Sea and Baltic Sea. In: Ozsoy & Mikaelean (eds.): Sensitivity to change: Black Sea, Baltic Sea and North Sea, NATO ASI Series, 2, Environment-Vol.27, Kluver Academic Publishers (pp: 516): 65-77.

Polonsky, A., Voskresenskaya E., Belokopytov, V. (1997) Variability of northwestern Black Sea Hydrography and the river discharges as part of global ocean-atmosphere fluctuations. In: Ozsoy & Mikaelean (eds.): Sensitivity to change: Black Sea, Baltic Sea and North Sea, NATO ASI Series, 2, Environment-Vol.27, Kluver Academic Publishers (pp: 516): 11-24.

Schulman, G.E. (1997) Effects of anthropogenic changes on the physiology of Black Sea fish, NATO-ARW: Environmental Degradation of the Black Sea: Challenges and Remedies, Constanta, Romania, October 6-10, 1997.

Vinogradov, M.Y., Shushkina, E.A., Musayeva, E.I., Sorokin, P.Y. (1989) A newly acclimated species in the Black Sea: The ctenophore *Mnemiopsis leidyi* (Ctenophora: Lobata), Oceanology, 29(2): 220-224.

Vinogradov, M.Y., Sapozhnikov, V.V., Shushkina, E.A. (1992) The Black Sea ecosystem, Moskva Russia Nauka: 112pp.

Zaitsev, Y. P. (1992) Recent changes in the trophic structure of the Black Sea. Fisheries Oceanography, 1(2): 180-188.

The Function of Wetlands

Boris Alexandrov

The largest Black Sea wetlands are situated in the coastal plain areas of Romania, Ukraine and the Russian Federation, in the deltas of large rivers with vast catchment areas: the Danube, the Dniester, the Dnieper, the Don and the Kuban. One of the principal wetlands in the area is the Danube Delta.

The word 'wetlands' spread globally in 1975 when the Ramsar Convention, the world convention for nature protection, proposed the definition of wetlands as habitats of waterfowl, areas of fen, marsh, peatland or water of either natural or artificial origin, constant or temporary, standing or drainage, freshwater, brackish or saline. The accumulation of gregarious birds in these zones is not random. It is the result of an 'edge effect' meaning life concentration on the border of coexistence of two or more ecosystems, terrestrial and aquatic for example or marine and riverine. As a rule there is higher productivity and biological diversity of plants and animals in these zones.

The Danube is the main river in the Black Sea basin and has a considerable influence on its productivity and biodiversity. About 50% of the total volume of fresh water entering the Black Sea comes from rivers of the north-western part, with 36% from the Danube.

The Danube Delta is the widest and richest wet zone in Europe after the Volga delta, the delta and the banks of the Black Sea amounting to 679,000 hectares. The Danube Delta has a classic form, being a very flat triangle with its apex at the point of division of the three branches: Kiliya, Sulina and Sfintu Gheorghe which transport 60%, 21% and 19% respectively at low river discharges. Although more than 78% of its area is situated in Romania, the Kiliya branch forms the newest and most productive part of the delta on Ukrainian territory.

The main physical and chemical factors causing the natural resources and high productivity of the Danube Delta are:

- a high level of irrigation: water and permanently flooded areas constitute about 80% of the Danube Delta, lakes cover not less than 10% of the delta surface and the drainage network is 1km/km^2;
- a high content of nutrients: study has revealed that some 58% of the total dissolved nitrogen - 340,000 tons - and 66% of the total dissolved phosphorus - 55,000 tons - flows into the Black Sea from the Danube basin;
- favourable temperature conditions: the average annual temperature is 12.7^{0}C; the average duration of water temperature at 5^{0}C, is 265 days.

These factors define the extraordinary biological status of the Danube Delta. The reedbeds, *Phragmites australis*, are the largest closed unit of reeds in the world, covering an area of 284,000 ha. The total intensity of organic matter production in the Danube Delta is 1.8 times higher than in the Dnieper Delta. The total number of species is above 4,000 including 1,120 aquatic organisms. In comparison with the Dniester and Dnieper Deltas the species composition is 1.7 and 1.2 times higher respectively.

Functions of the Danube Delta

The Danube Delta is a transition zone for aquatic, terrestrial, freshwater and marine communities and acts as a centre for condensing life in the Black Sea basin.

The terrestrial ecosystem

The richness and the high productivity of terrestrial systems of the Danube Delta can be judged by the state of the organisms which represent the top of the trophic pyramid of delta habitats, such as birds. There are 320 bird species in the Danube Delta, a considerably greater number of breeding bird species than in two other large southern European deltas, the Rhone and Guadalquivir. There are large numbers of wetland birds and a substantial proportion of the European, Palearctic or world populations of several species can be found here.

Of special note is the great importance of the Danube Delta for the pygmy cormorant *Phalacrocorax pygmeus*, the red-breasted goose, *Branta ruficollis* (over one tenth of the world population, 275,000, of this bird winter here), the white pelican, *Pelicanus onocrotalus* (the delta is one of the main Palearctic breeding sites for this species), the Dalmation pelican and the white-tailed eagle, *Haliacetus albicilla*, with eight pairs of the species in the Romanian area and three in the Ukrainian.

There are also significant populations of several species which are not yet in danger of complete extinction, but whose populations have declined drastically, especially in Europe. Of special note are the populations of colonial waterfowls, the glossy ibis, for example. The delta area is also a major wintering area for ducks and geese and it is also essential for migratory birds of northern Europe which fly over the area in tens of thousands on their way to Africa. Birds from Africa come to the delta in the summer to breed.

The delta is an important refuge for the European mink *Mustela lutreola*, the wild cat *Felis sylvestris* and the European otter, *Lutra lutra*. In 1992, 1994 and 1995, after an almost 50 year hiatus, the monk seal, *Monachus monachus,* has again appeared in the Ukrainian part of the Danube Delta. The Black Sea population of this species is about to become extinct.

The aquatic ecosystem

The Danube river-bed is enriched by plants and animals that live and develop in the river, first of all in the pelagic zone. The total volume of the plankton flow, besides nutrients, determines the efficiency of the mouth of the delta and the scale of influence on the Black Sea. The average annual volume of this flow at the top of the delta is about 1,340,000 tons, of which bacteria make up 80.8%, phytoplankton 11.1% and zooplankton 8.1%.

The zone of direct influence of Danube waters on the Black Sea is selected on the boundary of detection of freshwater algae which continue to grow in marine water. Depending on the estimate of the river runoff, the area of the zone varies and the maximum size of the surface of the pelagic zone reaches 100,000 km^2.

The increase in diversity, density and biomass of hydrobionts in the zone of transformation in comparison with adjacent areas is a manifestation of 'edge effect' on the boundary of coexistence of brackish water and marine fauna. In this zone, total biomass and production of hydrobionts are usually higher by two to five times. Regular blooming of the sea in a surface layer up to a depth of ten metres has been noted. The total phytoplankton biomass is more than 400,000 tons in an area of about 40,000 km^2 in the summer time.

Among the animal population of the ecotone 'river-sea' there is a prevalence of noctiluca, *Noctiluca scintillans*, making up to 90% of the density and biomass of pelagic organisms. In 1988 to the south of the Sfintu Gheorghe branch in an area of about 3,400 km^2 a super high biomass of this organism (125-560 kg.m^{-3}) was registered.

On average, the biomass of hydrobionts is five to ten times higher in the sea than in the river (phytoplankton - 4.8 times, mesozooplankton - 14.3 times, macrozoobenthos - 8.1 times) and lower numbers and biomass of hydrobionts have been observed in delta water in comparison with adjacent zones such as river branches and the sea-coast. The main reasons for this type of distribution of aquatic organisms is the intensive sedimentation (or silting) and reduction of current in the river delta.

The fish fauna of the delta is remarkably rich, with 91 species belonging to 30 families. Forty-four of these are freshwater species, the other being migratory species that occur in the Black Sea and mainly come to the delta during the breeding season.

After the construction of the Kahovka dam on the Dnieper river (1955-58), the Danube became one of the last rivers where sturgeons (starred sturgeon, *Acipenser stellatus*, and great sturgeon, *Huso huso*) continued to spawn.

The Danube Delta as a powerful natural biofilter

One of the most important functions of the delta for the Black Sea is its ability to absorb significant quantities of pollutants. There are a number of ways in which this is done:

- Absorption of pollutants on particles of organic substances. This process is carried out on particles of silt or detritus, taking them out from the water into the bottom sediments during the formation of the front edge of the delta. In this area each kilogram of dry weight of the bottom sediments contains on average two grammes of oil.
- The destruction of pollutants as a result of the functional activity of hydrobionts. In this way destruction and mineralisation of organic substances may be achieved. The microorganisms are the most active in this process (68%). The annual volume of organic matter destroyed by hydrobionts makes up about 1.5 millions tons.
- Absorption of pollutants by higher water plants. The dominating component of the ecosystem of the Danube Delta is higher water and coastal vegetation, which influences the water quality. About 93-99% of the total absorbed pollutants from the Danube accumulates in reedbeds.

The delta functions like the kidneys in filtering capacity in the body. The clearance rate for the main nutrients causing eutrophication in the Danube Delta is 39.7 kg per min for nitrates and 49.6 kg per min for phosphates. The functional productivity of the delta higher plants can be estimated over an annual period as: nitrates - 59,100 tons, phosphates - 20,500 tons, heavy metals - 23,300 tons and about 100 kg pesticides. The intensity of absorption of pollution in the Danube Delta is 13 times higher than in the Dniester and 16 times higher than in the Dnieper.

The Danube Delta as a region of human economic activity

From ancient times man has used the natural riches of the Danube Delta. The first signs of permanent human settlements in the delta go back to the Neolithic epoch, with a centre at Hamangia - now Baia, Romania. The Ukrainian part of the Danube Delta is quite young - about 300 years old.

Today, in spite of its vast natural riches, the density of the population of the Danube Delta is small, with 14 people per square kilometre and a total population of 83,600. The main types of economic activity in the Danube Delta region are:

- Fishing (37.0 %). Fishing in the branches of the Danube has always been closely related to the flood zones upstream and there are only a few species: migratory species (sturgeons - the Russian sturgeon, *Acipenser guldenstadti*, the starred sturgeon, *Acipenser stellatus,* and the great sturgeon, *Huso huso,* and shad, *Alosa kessleri pontica)* and semi-migratory species (carp, *Cyprinus carpio*, zander, *Stizostedion lucioperca*, bream, *Abramis brama,* and eels, *Siluris glanis*).

The annual volume of the fish catch in the Danube Delta for the 1981-1990 period was about 2,020 tons: 59% of this is wild goldfish, *Carassius auratus*, 38% is shad, 1% is sturgeons. Distribution of the total fish catch on average is about 70% in the Romanian part, and not more then 30% in the Ukrainian part. Today, the spring catch of migratory fish may employ up to 500 fishermen in Romania and 600 fishermen in the Ukraine. The first fish polders were constructed in the 1950s and intensive fish farm production was 338 kg/ha in 1986.

- Reed exploitation (19.9 %). The potential mean yield of natural reeds is 6 tons/ha. The modern level of reed exploitation is not more then 50,000 tons per year. This is about 6% of the possible volumes. This kind of economic activity is widespread in Romania.
- Agriculture (25.9 %). Agricultural land covers about 139,600 hectares with an equal part in Romania and the Ukraine; the polder surface area altogether being about 36,000 hectares. This is about 24% and 8% of the delta area respectively.
- Navigation. In 1989 boats carried about 30 million tons of cargo along the Danube Delta branches, two-thirds of which was from the Black Sea.

Damage to the Danube Delta

Several interrelated factors are responsible for the damage to the functioning of the Danube Delta:

- The high nutrient loads of the Danube River upstream of the delta. Data provided by the research institutes reveal that the delta is subject to eutrophication which during the summer can lead to massive fish mortalities caused by dissolved oxygen deficiency and the release of hydrogen sulphide; another effect of eutrophication is the disappearance of submerged macrophytes which are indispensable for the breeding of certain fish species, pike for example.
- Engineering impoundments, dams, dikes, cut-offs, canalisation, which break the connection between the river and its floodplain. For example, from 1963 to 1967, embankments on the Romanian stretch of the Danube river reduced the area of the floodplain by 290,000 ha, which corresponds to a loss of 4.3 km^3 of flood retention capacity.
- Intensive agricultural practices and fish farming. The creation of polders has reduced the flooded area by 25% since 1960 and the water retention capacities of the delta

have been reduced to 1.7 billion m^3. There has been a decrease in predatory fish to the benefit of non-predators, and a replacement of carp by goldfish.

- Changes in the hydraulic regime of the Danube River. Over the last few years this has decreased the area of flooded land available for the successful reproduction and healthy growth of the young fish.

Looking to the future

In recognition of the significance of the Danube Delta, it was proposed that UNESCO declare the area part of the heritage of humankind. The Biosphere Reserve in the Romanian part of the delta was created on 27 August, 1990, and includes the lake complex and spans an area of the Black Sea extending from the shore to 20 metres in depth. In the Ukrainian part of the delta there is a plan to create a Biosphere Reserve with a total area of 46,403 ha, based on the existing nature reserve, Dunaiskie Plavni.

The setting up of a joint environmental complex in the Danube Delta was proposed by the official representatives of the Man and Biosphere Programme from Romania and the Ukraine at an international forum in Seville in March 1995 at which 324 biosphere reserves from 82 countries around the world were represented. This was supported by the International Committee. The GEF has recently assisted Romania and Ukraine with a World Bank administered grant of some US$ 5 million to improve management of the two parts of the reserve.

With all possible measures being undertaken to ensure that the Danube Delta is protected, it is to be hoped that the saying 'we never value what we have, but cry when we lose it' never comes true for the Danube Delta.

Chapter 4

Degradation and Conflict

Introduction

Charles N. Ehler

About 60% of the people in the world now live within 100 kilometres of a coastline. Within the next three decades, 75% of the population could reside in coastal areas. The population of the world doubled between 1950 and 1985 and today it is growing at a rate that adds a billion people every 11 years. The next decade might represent the last chance to stabilise human population at something less than double the current world population of 5.7 billion by the middle of the next century. The next doubling of the world's population will be of far greater significance in terms of energy, resource consumption and stress on the environment, especially the coastal environment, than any previous doubling of world-wide population. Today's vague concept of 'sustainability' will take on very real proportions as we move into the next millennium.

The roughly five-fold increase in the number of human beings over the past century and a half is the most dramatic terrestrial event since the retreat of the ice-age glaciers thousands of years ago. That explosion of human numbers has been combined with an increase of about four-fold in consumption per person and the adoption of a wide array of technologies that needlessly damage environment. The result has been something like a twenty-fold escalation since 1850 of the pressure humanity places on its environment - an unprecedented assault on natural ecosystems.

Historically, when coastal populations were low, human impacts on coastal areas were also low. Perhaps more importantly, the impact of one person's activities had little effect upon his neighbour. Unfortunately, today's reality is different. Coastal populations have now reached the point where one kind of coastal use conflicts with another, simply through lack of space. All too often, one person's solution becomes another person's problem.

The fingerprints of humanity are found everywhere in the sea. Chemical contamination and litter can be observed from the poles to the tropics and from beaches to abyssal depths. However, conditions in the marine environment vary widely. The open ocean is still relatively clean. In contrast, coastal areas and enclosed seas such as the Black Sea are affected by humans almost everywhere. Increasing human stresses on coastal areas can be observed world-wide. If unchecked, this trend will lead to further deterioration in the quality and productivity of marine and coastal environments.

On a global basis, more than 80% of the pollution in the sea comes from sources on the land. This situation is no different in the Black Sea. Pollutant discharges from industry and sewage treatment facilities, together with runoff from agriculture, forestry, and urban areas within the coastal zone are major sources of marine pollution. However, massive quantities of pollution from sources far upstream are carried into the coastal zone and then into the sea by major rivers. Almost two-thirds of the nutrients flowing into the Black Sea comes from the Danube River basin.

Experience throughout the world demonstrates that economic development often does not show sufficient concern for maintaining natural systems and environmental quality. This is due in part to the view that economic development and environmental quality are alternatives; deterioration of environmental quality is viewed as a necessary cost of economic development. This view is misleading.

For example, pollution of coastal waters can destroy commercial or recreational fisheries and check economic development related to both fisheries and coastal recreation and tourism. Water pollution can affect human health with a resultant loss in overall productive effort as well as direct welfare losses to individuals through, for example, water pollution-related illnesses.

In the first paper of this chapter, Arabinda Basu explains what Hinduism has to say about ecology and the environment. He argues that the part that we play in destroying or improving the environment depends on our state of 'consciousness', which includes not only awareness, but dynamic and creative energy as well. He pleads for a basic change of consciousness so that we can review our concept of good, our view of nature and our responsibility for the environment around us. The time has come to cooperate with nature so that she can give us as much as she can and man in turn must not extort more than she can yield at any time. Paraphrasing Sri Aurobindo, he concludes that 'one who loves man but does not love Nature, does not love man either'.

Radu Mihnea describes pollution problems and sources in the Black Sea. He points out that land-based sources are the most important, including pollution carried to the Black Sea by major rivers, such as the Danube. While chemicals, radionuclides and sewage represent serious problems in the Black Sea, eutrophication (from the Greek, meaning 'well-nourished') is by far the most damaging to the natural environment. Excess nutrients from agricultural wastes, fertilisers and sewage enter coastal waters and cause dense blooms of phytoplankton, which in turn decrease water transparency and light penetration. Without light the plants begin to die and decompose, consuming oxygen in the process. Severely-reduced levels of oxygen in the water column often leads to the death of marine organisms with important economic consequences.

The two final papers argue for improved coastal management of coastal areas in the Black Sea. Using examples of conflicts in the development of the Russian segment of the Black Sea coast, Erdal Özhan emphasises the importance of managing conflicts from a long-term perspective and the consideration of the outcomes of today's decisions on future generations. For example, Gelendzhik, once a prosperous tourist and wine growing centre, faces a difficult future because of massive harbour dredging and severe pollution from agricultural and municipal wastes that have been directly discharged into the Black Sea. Sylvie Goyet points out that integrated coastal management, often understood as a government-led process, can in fact be initiated and carried out by local communities. She argues for increased participation in the coastal management decisions by all stakeholders, especially for increased representation of local inhabitants most affected by management decisions. Without wide representation and effective participation, most plans are, in the long run, never accepted or implemented.

The challenge is to balance short-term development needs against the long-term sustainability of ecosystems, habitats and resources so that the range of choice and opportunities available to future generations is not diminished by the consequences of development choices today. That challenge will not be met until fundamental changes are made in human consciousness and values.

The Sacredness of all Living Organisms

Arabinda Basu

It is my task is to expound briefly what religion has to say about ecology and the environment from the Indian point of view. By religion I do not mean a system of thought, a creed or a cult, rituals and ceremonies or prohibition and injunctions regarding human behaviour. All these things are part of all the great religions and are useful elements of them, but they are not the essence of religion.

> The deepest heart, the inmost essence of religion, apart from its outward machinery of creed, cult, ceremony and symbol, is the search for God and the finding of God. Its aspiration is to discover the Infinite, the Absolute, the One, the Divine, who is all these things and yet no abstraction but a Being. Its work is a sincere living out of the true and intimate relations between man and God, relations of unity, relations of difference, relations of an illuminated knowledge, an ecstatic love and delight, an absolute surrender and service, a casting of every part of our existence out of its normal status into an uprush of man towards the Divine, and a descent of the Divine into man.[1]

It may seem that this idea of religion emphasises man's relation to God, but does not stress his relation to his fellow-beings. That aspect of religion is clearly brought out in another pregnant sentence of Sri Aurobindo: he has also written that one who loves God but does not love man does not love God either. If this be the core of religion, then what does it have to do with ecology and the environment?

To find an answer to this question we must enquire what Indian spiritual thought has to say about God and Nature. The briefest and the most comprehensive description of God is found in the Vedanta, the most developed and effective system in Indian spiritual thought and practice. God is *Sat*, *Chit* and *Ananda*, Existence, Consciousness and Bliss or Delight. These words suggest an abstract impersonal Reality. While that is one aspect of God, the Vedanta of the Upanishads also affirm that God is personal, that he is Existent, Conscious and Enjoyer of the Bliss of Self Conscious Self Existence.

Consciousness in this description of God is not

> a phenomenon dependent on the reactions of personality to the forces of Nature and amounting to no more than a seeing or interpretation of these reactions. The subjective personality itself is only a formation of consciousness which is a power inherent, not in the activity of the temporary manifested personality but in the being, the Self or Purusha...

> Consciousness is not only power of awareness of self and things, it is or has also a dynamic and creative energy. It can determine its own reactions or abstains from reactions; it can not only answer to forces, but create or put out from itself forces. Consciousness is Chit but also Chit-Shakti [Conscious Force].[2]

God does not need religion but man, as a conscious being, does. God is Reason according to Christian and Islamic theology, but surely not human reason. God does not have to know or act with the help of the often deceptive senses and the very limited capacities of the human mind. I would suggest that what Vedanta means by Consciousness, *chit*, is what the Semitic religions describe as the Reason of God, or God as Reason, the Word of John's Gospel. My point is that the part man plays in destroying or improving the environment depends on his state of consciousness.

According to Hinduism the real man, the true self of man, is of the same essence as God. The word *dharma* is derived from the root *dhr* (*dhri*), which means to hold, to support. Hindus should conduct their lives according to *dharma*, which is the Law of the universe and thus of human life, both individually and collectively. The term has moral and ethical, juridical, religious and spiritual significance, all of which are connected. For all practical purposes it means religious morality, which according to one system of Hindu philosophy leads to both worldly prosperity and spiritual liberation. Those in India who were atheists also observed *dharma*, the principles of individual and social morality and ethics. In the context of discussions on devising practical measures for saving and improving the environment, it is my opinion that the moral aspect of *dharma* is essential.

Many Jewish, Christian and Muslim writers say that the world was created by God, and therefore man is expected to revere Nature. Vedanta not only says that Nature and the world are entirely dependent on God but that He has become the world. He is both the instrumental and the material cause of the world. This may sound like pantheism but Vedanta holds that God is always both transcendent of and immanent in the universe which is nonetheless God in becoming. The root of the word in Sanskrit for the becoming of the world is *Srj,* which means to loose forth, to bring out into open manifestation what is contained within. If this be so, how much more should man respect, revere and love Nature, how much more should he care for and tend our planet, because that would be the means of fulfilling God in the world?

Since the nineteenth century, man's God has been Mammon, and his Goddess, Success. The consequence is that since then everything has been grist to the mill of sensation and enjoyment. In this process the environment has become a hapless victim.

In ancient India, human life was divided into four spans; *brahmacharya, garhasthya, vanaprastha* and *sannyasa*. These are, respectively, the stage of studentship in which study and moral training were the main duties of young people; the time to marry, earn money, work for prosperity and raise a family; after retiring from the practical business of life, a time to concentrate on developing a keen aspiration for spiritual liberation; and finally, when minds and hearts were sufficiently pure and worldly ties or desires did not bind, the stage of renouncing all mundane values to practise spiritual disciplines in order to achieve spiritual liberation.

The moral foundation for the stage of the householder was laid in the first stage. The students lived with their gurus, their moral, intellectual and spiritual guides, in the forests. They developed respect and love for Nature and were trained to look after its welfare throughout their life. The householder was instructed to earn enough to support himself and his family comfortably. After this obligation had been met, he was instructed to dig wells and ponds, plant trees, build shelters for travellers and places where drinking water could be given to them, activities meant to preserve Nature and enhance its life and beauty. The fulfilment of these duties earned the householder religious merit and thus economic prosperity, moral duties and religious merit were all were rolled into one concept and practice, that of *Dharma.*

Today the situation is sadly very different and India is facing a great cultural problem. People chase relentlessly after economic prosperity and social success. Trees are cut down causing an acute shortage of drinking water; the building of dams helps industrial development but also inundates many villages. The villagers become homeless and many flock to nearby cities and towns, this in turn causing environmental problems. The increasing sales of cars and motorcycles helps the economic prosperity of the country but adds daily to pollution in the cities and their suburbs.

India is a poor country and people have a right to improve their lot and raise their standard of living. However, when people start thinking that better living is the be-all and end-all of their existence and the ideal of a good life is lost sight of, then things start to go wrong. The problem is, how can we bring about a reconciliation of the idea of good living and the ideal of a good life?

In ancient India there was no conflict between science and religion. India made great contributions to mathematics, astronomy, chemistry, medicine and psychology. Some of the scientists were also spiritual sages and saints. It was recognised that the natural world and the realm of the spirit were two different areas of enquiry with respective means and methods of investigation. It seems to me that modern conflict is not so much between science and religion as between the scientific view of the world and the religious perspective of it. The latter did not, in ancient India, deny the findings of science about the natural world though it did claim that the truths of science were not complete. Science now has no right to dictate thinkers' methods of attaining knowledge to the religious, nor should religious leaders deny the truth of the sciences.

The question boils down to man's sense of value which is a matter of consciousness. Scientific discoveries are in themselves neutral. The use made of them is decided by men. Atomic energy can be utilised for electrification and irrigation or to make atom bombs. I am pleading for a basic change of consciousness in man so that he can review his concept of good, his view of nature and his responsibility to the environment around him. Those who are concerned about improving the environment must take a stand against those scientific discoveries and their practical applications which affect the environment unfavourably. Unfortunately, political rivalry, military expediency, interests of economic imperialism and the struggle for markets seem more important than the preservation of the environment and the improvement of the quality of life. There is a steady corrosion of our sense of values. It is imperative that we cleanse our sense of priority of our needs and wants. They are by no means the same.

Auguste Comte proposed a Religion of Humanity.[3] What he meant, I think, is that we should cherish humanity and do all in our power to improve the quality of mankind's existence. It seems to me that today we need a Religion of Nature. Refined sentiment, respect, reverence and even love must be cultivated if we would not exploit Nature in our inordinate clamour for exciting and sensational pleasures. The time has come to cooperate with nature so that she can give us as much as she can and man in his turn must not extort more than she can yield at any time. Ancient Hindu thought described man as a child of the Earth, which was the mother, and of Heaven, the father. Even if not everyone can believe in God, we can at least revere and love and do all we can to preserve the health and beauty of Earth and Heaven, our 'parents'.

Echoing Sri Aurobindo's profound and beautiful remark that one who loves God but does not love man, does not love God either, we may well say that one who loves man but does not love Nature, does not love man either. Let us salute Nature with humility, gratitude and love.

Notes

1. *The Human Cycle.* In *Social and Political Thought*, Vol.15, p.122.

2. *Letters on Yoga*, Sri Aurobindo Birth Centenary Library Vol 22, pp.233-34.

3. *Considerations on the Spiritual Power*, (1826). Auguste Comte actually substituted the Great Being for God - humanity of the past, present and future, but the phrase 'religion of humanity' has been inevitably associated with the name of Comte.

Pollution Problems and Sources in the Black Sea

Radu Mihnea

The most important polluting factors in the Black Sea are land-based sources which account for more than 70%. These sources are tributary rivers (either national or international) and point sources (domestic and industrial discharges). The dumping of different materials, including dredge spoils, is also a contributory factor.

River transport is by far the most important source. In 1995 a region-wide assessment of the land-based sources of pollution was performed by one of the former Activity Centres of the Black Sea Environmental Programme. According to this and to the Transboundary Diagnostic Analysis performed also in the BSEP, of the total of 1,035,635 tons of Biological Oygen Demand (BOD) introduced into the Black Sea per year, 882,257 tons (97%) come via rivers. For Total Suspended Solids, of 53,976,963 tons 99% belongs to rivers. In the case of total pollution, the annual quantity of 51,726 tons per year is composed of 42,601 ton (83%) from rivers and only 8,675 ton from land-based sources.

The most important rivers from the point of view of pollution are the north-western ones - the Danube, Dnieper and Dniester. The Danube counts for 87.8% of BOD, 70.7% TSS, 94% total nitrogen and only 35% total phosphorus for riverine transport. Annually, the Danube alone discharges into the Black Sea about 1,000 tons of chromium, 900 tons of copper, up to 60 tons of mercury, 4,500 tons of lead, 6,000 ton of zinc and 50,000 tons of oil.

The most important environmental problems in the Black Sea are chemical pollution, eutrophication, and changes in marine ecosystems.

Chemical pollution

A large amount of data has been gathered on the heavy metals content of the Black Sea. The data comes from different laboratories in the coastal states or from international programmes such as COMSBLACK, EROS 2000 and NATO. Some of them have been analysed by means of normalising parameters (aluminium and organic content) some have not. The conclusion is that the trace metal contamination is generally limited to areas where high concentrations occur, well defined source-related hot-spots, and is not a basin-wide phenomenon. Those hot-spots are in front of deltas or estuaries, commercial ports or industrial water discharges. Unfortunately these are the coastal waters where the most important biological processes take place.

Concentration of pesticides and polychlorinated biphenyls are relatively low. Non-organophosphorus compounds were determined in marine organisms in an analysis performed in 1991-1993 in Romania and some higher concentrations of organochlorine compounds were found near the Danube Delta and Sinoe lagoon. In order to assess the effects of heavy metals and chlorinated pesticides a 'mussel watch' programme has been developed.

Oil pollution

It was estimated that 111,000 tons of oil enter the Black Sea each year. Of this quantity, the Danube accounts for 48%. Most of the remainder comes from land-based sources (wastewater without proper treatment and accidents in handling oil or oil products). A considerable amount of oil reaches the Black Sea from ballast water discharged by ships but an accurate figure is unknown. A report by a Russian scientist published in 1985 shows high levels of oil products along the navigation route from the Danube or north-western part of the Black Sea to the Bosphorus. There are inadequate port reception facilities in the Black Sea and deballasting activities are reported to occur especially in the north-western part where Economic Exclusive Zone boundaries have not yet been agreed upon.

Radionuclides

The concentration of artificial radionuclides in various components of the Black Sea environment is well studied by scientists from all coastal states as well as from other western countries. As a result of the accident at Chernobyl there was an increase in the values of some radionuclides (Cs-137, Cs-134 and Sr 90) but they remained within acceptable limits for human intake and it is estimated that the Cs-137 will be normalised in the Black Sea surface waters by the year 2000.

Sewage

Sewage is an important source of organic substances (dissolved or particulate), heavy metals, persistent organic pollutants, detergents and pathogenic bacteria. The principal sources of sewage contamination of the Black Sea coastal are rivers contaminated by up-stream sewage discharges and direct discharges as well as direct discharges from coastal settlements into the sea.

The total waste water discharges from communities in the Black Sea coastal area, estimated at 55m^3 per year per resident, is:

Country	*Sewage discharge m^3/year*
Bulgaria	36,300,000
Georgia	26,675,000
Romania	22,000,000
Russia	44,000,000
Turkey	237,600,000
Ukraine	204,600,000
Total	571,175,000

144,572,500 m^3 untreated sewage are discharged in the Black Sea annually, 88,385,000 in the shallow north-west shelf where dilution effects are less and where there is extensive tourism.

Eutrophication

Anthropogenic eutrophication of coastal waters has been the most damaging of all the many harmful human influences on the oceans and seas, both in terms of its scale and its consequences.

During the last three decades, among other man-induced changes, increased eutrophication has considerably changed the structure and functioning of the Black Sea ecosystems, mainly in the north-western corner.

There are sufficient historical data to demonstrate that there has been an enormous increase in nutrient load to the Black Sea in the past 25 years. In the 1950s the Danube, the Dniester and the Dnieper annually carried 14,000 tons of phosphates, 155,000 tons of nitrates and 2,350,000 tons of organic matter into the north-western Black Sea. By the 1980s the rivers were carrying an annual average of 55,000 tons of phosphate (about 4 times more), 340,000 tons of nitrates (about 2 times more) and 10,700,000 tons of organic matter (about 4.5 times more).

Recent estimates for the Danube basin and its tributaries have revealed that about half the nutrient load comes from agriculture, a quarter from industry and a similar amount from domestic sources.

For the Romanian coastline, which is under the direct influence of the Danube, the larger quantities of nutrients represented an increase of more than 20 times for phosphorus at the last period of the 1980s compared with the beginning of the 1960s and about five times for nitrates.

The immediate consequences of increased nutrient supply was the intensification of phytoplankton blooms, particularly on the shallow north-western continental shelf. Under the new conditions the development of almost monospecific, dense blooms of phytoplankton (especially Dinoflagellates) was encouraged and between 1981 and 1990, 46 algal blooms produced by 15 species were registered in Romanian littoral waters. For this period, *Prorocentrum cordatum* reached the enormous density of 807.6 million cells l^{-1} compared to 50.8 million cells in the period from 1960 to 1970.

In the Romanian sector of the Black Sea, the average biomass of phytoplankton reached 4105 mg. m^{-3} from1983 to 1990, eight times higher than the average of 495 mg. m^{-3} recorded in the period from 1959 to1963. For the Odessa bay the increase was even more important: biomass increased 26 times in 30 years. For the whole of the Black Sea, maximum values of phytoplankton increased from 52 to 800 mg.m^3 during the last three decades.

Such phenomena were initially beneficial to the marine food chain but the longer term effects have proved to be devastating. There has been a severe decrease of water transparency; Secchi depth measurements have decreased from 50-60 metres in the early 1960s to values of 15-16 metres in the open sea or a maximum 3-5 metres in coastal waters (20-30m isobate).

As a direct consequence of decreased light penetration, a massive loss of macrophytic algae has been registered. The only algae adapted to dim sunlight survived at depths of over 15-20 metres. Beside limiting the species composition and reducing the biodiversity it has to be considered that some of those algae represented a major economic resource. This is the case of Zernov's field of *Phyllophora* in the central part of the Black Sea. In the 1950s *Phyllophora* covered an area equal to 10,000km^2 with a total biomass reaching about 10,000,000 tons. Towards 1980, it was reduced to 3,000 km^2 in area and 1,400,000 tons in biomass. In 1990, these values had declined to 50km^2 and 300,000 to 500,000 tons.

Apart from its economical value (it was harvested for the agaroids), *Phyllophora* was an important source of oxygen. According to some data (Zaitsev), it produced up to 12,000,000 m^3 of oxygen per day. The associated fauna of the *Phyllophora* field disappeared, especially sponges, sea anemones, isopods, shrimps, crabs plus more than 40 species of fish. *Phyllophora* biocoenocis consisted of 118 species of invertebrates and 47 species of fish with special adaptations - protective red colour - to match the colour of *Phyllophora*. With the disappearance of *Phyllophora* the entire biocoenosis collapsed.

Another algal representative, the brown alga *Cystoseira barbata*, began disappearing along Romanian and Ukrainian coasts in the 1980s. This large perennial alga, the most important one on the rocky coasts, cannot endure eutrophic marine waters and was replaced by filamentous green algae. Cystoseira used to be the specific and support organism for a typical bicoenosis composed of dozens of species of coastal invertebrates and fish.

The exuberant growth of phytoplankton during the last decades has had serious consequences on the zoobenthos. Algal blooms generate enormous quantities of dead organic matter in the water column. As a result, because of the amount of oxygen consumed during its decomposition on the shelf regions, hypoxia and anoxia began to appear. This was first observed in the Black Sea in 1973. After this year, oxygen depleted zones on the north-western corner were noted every year in summer-autumn seasons, their scale, duration and consequences increasing year by year. For the benthos, one of the most damaging effects of eutrophication is the creation of those zones in the bottom layers with reduced or no oxygen.

Different species of bottom organisms react differently to low oxygen concentrations. For most of them an oxygen concentration below 1.5-2 ml per litre is lethal. Only some polychaetes are more tolerant to hypoxia being capable of living for some time in water with oxygen concentration below 0.5 ml per litre. Bottom and near-bottom invertebrates and fishes fell victim to hypoxic conditions, especially those that were not very fast swimmers. The qualitative and quantitative structure of benthic communities has changed, sensitive populations have diminished while those resistant to alteration of the environment conditions have proliferated. On the Romanian coast, the macrofauna from the sandy bottoms consisted of 69 species at the beginning of the 1970s; in the middle of the 1980s only 14 species were identified.

In some years, the area depleted in oxygen reached 30,000 to 40,000 km^2. The biological losses due to hypoxia and anoxia on the north-western corner are estimated at 100 to 200 tons of animals per square kilometre of seabed. The total biological losses over 18 years (1973-1990) are estimated at 60 million tons, including 5 million tons of fish (Zaitsev). Very often in summer, in the shallow water near the shore, huge quantities of dead or near dead fish and invertebrates are thrown out on the beach by the waves. On the Romanian coastline alone, a single event of anoxia in 1991 eliminated an estimated 50% of the remaining benthic fish. The same thing had happened to molluscs. The total losses induced by anthopogenic influences (hypoxia and silting) across the 3,300km^2 seafloor in this area has been estimated at 800,000 tons of biomass, the majority of which were mussels.

The death of vast quantities of animals in the benthic zone does not just mean the disappearance of valuable commercial resources with resulting economic losses. The ecological damage is enormous including the consequences of mass mortalities of mussels and other filter-feeding organisms. Research into filtration activity of mussels has shown that 1 km^2 of mussel biocoenosis can filter an average of 15-20 million m^3 of sea water every 24 hours. In the 1960s, mussel biocoenosis on the north-western shelf covered an area of about 10,000 km^2. It is difficult to assess the ecological impact of the loss of such a huge number of active filter-feeding organisms in conditions where the amount of phytoplankton, bacterioplancton and detritus has increased in such a manner, but it must have been significant and for some years serious efforts have been made in Romania to increase the lost natural biofilter with artificial reefs and intensive mussel cultures.

Another change in the Black Sea ecosystem which occurred as an effect of eutrophication and has important biological and economical consequences is the introduction of new species.The most spectacular although unpleasant one was the exotic ctenophore *Mnemiopsis leidyi*. Mnemiopsis feeds on planktonic crustaceans, mollusc larvae and pelagic fish eggs and larvae and its presence contributed to the sharp decline in anchovy stocks in the Azov and Black Sea. The first specimens were reported in 1982 and by the late 1980s it had reached a total biomass considered to be close to one billion tons.

Bibliography

Balkas, T., Dechev, G., Mihnea, R. Serbanescu, O., Unlata, U. (1990) *State of the marine environment in the Black Sea,* UNEP Regional Seas Reports and Studies, No 124.

Bologa, A.S., Bordeanu, N., Petran, A., Tiganus, V., Zaitsev, Y.P., (1995) *Major modification of the Black Sea benthic and planktonic biota in the last three decades.*

Bulletin de l'Institut océanographique, Monaco, No. special 15, pp. 85-110.

GEF Black Sea Environmental Programme (1996) *Analysis of Fleet Structure and Performance in the Black Sea Fisheries,* Draft Report.

GEF Black Sea Environmental Programme (1996) *Black Sea Transboundary Diagnostic Analysis.*

GEF Black Sea Environmental Programme (1996) *Region-wide assessment of the land based sources of pollution in the Black Sea*, The Activity Centre on Routine Pollution Monitoring.

Mee, D. Laurence *Management and protection of the Black Sea environment: an international approach,* Envir. Econ. Expert Group, 1992, pp. 1-13.

Zaitsev, Y. P. *Recent changes in the trophic structure of the Black Sea.* Fisheries Oceanography Vol 1 N. 2, 1992, pp. 180-189.

Zaitsev, Y. P. and Mamaev, V *Marine Biological Diversity in the Black Sea. A study of change and decline*, United Nations Publications,1997, p. 208.

Coastal Conflicts: the Need for Management

Erdal Özhan

The coastal areas of the Black Sea are beautiful and highly valuable. Steep and wild rocky slopes, linear and pocket beaches, internationally important wetlands and lagoons, attractive landscapes and sea views, fertile agricultural land, and the pleasant subtropical climate are among the features of these areas. The Black Sea coastal areas today are not yet heavily pressured by congestion from human populations and by intense development. However, inappropriate development in the past, poor infrastructure, and lack of planning and management have resulted in various hot-spots, mainly in developed enclosed areas like bays, which suffer from significant degradation of the environment and resources, and need urgent action for rehabilitation and restoration. The value given to the conservation of nature and environmental issues differs significantly among various sectors of society and when faced with the dilemma of economic activity versus nature and conservation, people's priorities vary considerably. Hence there is conflict.

Different types of coastal conflict can be illustrated by examining what has been happening on the Russian part of the Black Sea coast. The problems are not specific to this particular part of the coast, and can be seen elsewhere as well.

About 30 kilometres south of Novorossiysk is the town of Gelendzhik, with about 50,000 inhabitants, located on the shores of Gelendzhik Bay. This area was one of the important centres of the former Soviet Union for tourism and recreation, and was heavily subsidised by the state. During this time, Gelendzhik was one of the more prosperous towns, with major economic inputs from tourism and wine production. The beautifully shaped Gelendzhik Bay is a highly enclosed, shallow water body, naturally protected from the forces of the sea. The land around the bay is elevated, providing the advantage of attractive sea views from distant locations. A very attractive forest covers the land adjacent to the bay, and part of the town is built in the forest. Development, with the exception of two high rise hotels, is modest.

The shores of the Gelendzhik Bay do not naturally have sandy beaches and in the past, several coastal projects have been implemented for preparing the area for recreational activities. The northern part of the coast is heavily structured with groynes and revetments. The southern segment has been converted into a series of man-made beaches by the addition of sand and gravel. In one project carried out in 1970, the sand dredged from the middle of the bay was pumped ashore, creating a big cavity on the sea bottom. A large pipe buried underground collects wastewater from the city, and discharges the effluent, which also has agricultural runoff from the surrounding wine plantations, right into the bay in front of the city hall. There is a pier at the south-east end of the bay which supports a modest level of shipping activity.

Studies carried out by the Southern Branch of the Shirshov Institute of Oceanology in Gelendzhik indicate that the water and the bottom material in the bay are severely polluted. The recreational use of the bay is associated with health risks and infections through contact with the water in the summer season. The discharge of untreated waste water into the bay and the modest shipping activity over the years have introduced heavy metals (including mercury, cadmium and zinc), oil, and pesticides such as DDT, the use of which was banned internationally many years ago. The oil concentration at the bottom soil has a thickness of 40cm which peaks near the pier and is many times greater than that in the Tuapse Bay, which houses an oil terminal. The pollution of sediments in the trench created by the dredging operation is more severe than in the surrounding areas, due to the accumulation of pollutants in the deeper, man-made, part of the bay. There is now a plan to fill this artificially made trench with rubble as a measure against further accumulation of pollutants.

The example of Gelendzhik Bay, which is very similar to several other highly enclosed small bays of the Black Sea, vividly illustrates several types of coastal conflicts. Firstly, the finite space and finite resources of the bay put a limitation on the potential uses of the bay which should be carefully planned. Secondly, some coastal uses, such as tourism and recreation are incompatible with others, such as transportation or waste discharge. The digging of the trench in the middle of the bay and the current plan for filling it back with rubble clearly illustrates the importance of a scientific input into coastal development and management.

During the recent period of transition, Gelendzik has been undergoing an economical decline due to the decrease of tourism in the area. The majority of people who could visit Gelendzhik in the past through governmental subsidies are now unable to afford it and the minority - the new rich - prefer foreign destinations which offer a higher quality of services. There are no resources available to improve the infrastructure (there is a severe shortage of fresh water for household use in the summer months) and the waterfront on one hand, or to develop new accommodation facilities and other support services of reasonable quality on the other. Both are desperately needed to revitalise the tourism industry.

A private company has proposed to build a commercial port within the bay near the entrance at the northern shore. A recent referendum has revealed that about 90% of the population favours this project, hoping that it will provide them with jobs since the majority of the adult inhabitants either have no jobs or are very poorly paid. The local administration also appears to be in favour of the project, with the expectation that the tax revenues from the port will ease the financial bottleneck and lead to improvement of the infrastructure that is necessary for the revitalisation of tourism. Only a small group who can predict the future implications of the project oppose it. It is no doubt that the port project, if implemented, will have a dominant mark on the future development of the city and the bay of Gelendzhik. It is likely that the majority of the people who vote favourably for the project because of their immediate economical needs and insufficient environmental awareness, will regret this sometime in the future, as they will see their little town change into a messy industrial centre and the natural resources including the vineyards, the forest, and the very attractive land and seascape, disappear. This example clearly illustrates coastal conflicts which may result from alterations of the values and priorities of populations in time.

Man-made changes to the coastal systems, whether it be the cut-off of the sediment supply due to construction of dams on rivers, blocking of the littoral drift by the harbour facilities and moles, or interfering with the dynamic equilibrium of the beach and dune systems by building too closely to the shore, are among the causes of coastal erosion at many sites along the Black Sea coast. It has been well documented that the anticipated sea level rise will accelerate coastal erosion.

Pocket beaches, numerous along the Black Sea coast, are among the coastal systems which are usually in a delicate dynamic equilibrium, and respond unfavourably to any man-made changes. A good example is the small mole, built during the Second World War, for the berthing of military ships, which divides the beautiful, gravel pocket beach formed at the outlet of the River Aderba, in Divnomorskoye, south-east of Gelendzhik. As the material supplied by the river is intercepted by the mole, the beach has been widened at the northern sector of the mole at the expense of almost complete erosion of the southern part. The eroding pocket beach, being an important area for recreation and tourism, has received the attention of the authorities, who have spent substantial amounts of money building sea walls and groynes to stabilise the beach. When this failed, they built ugly, elevated concrete platforms parallel to the coast for the sunbathers. It is clear that there is an urgent need for scientific information on the consequences of man-made alterations to the coastal systems, and for managing erosion and rehabilitating the eroding areas.

The need for management to avoid incompatible and mutually harmful patterns of use is also very well illustrated by the present developments around the shores of the Cemessk Bay at Novorossiysk. Novorossiysk has a population of about 300,000 and is the largest Russian urban centre and port along the Black Sea coast. It occupies a part of Cemessk Bay, which has a very picturesque and attractive coast in spite of problems of water and air pollution and destruction of the landscape behind the harbour due to mining of raw material for a cement factory. Important uses of the area around the bay include industry (the cement factory, storage and transport facilities built at the start of the century), a sizeable commercial port, an oil terminal (a pier and oil storage facilities), urban use, and recent development for tourism at the southern part of the outer bay. The water inside the bay, which is less enclosed than Gelendzhik Bay, is heavily polluted by oil from tanker operations, discharges from the cement factory, and wastes from bulk cargo handling in the port, such as coal, and domestic discharges. Still the bay is used for the recreational needs of the citizens, and people fish in the bay, even in the heavily polluted harbour area. The present pattern of use of the Cemessk Bay clearly shows the low level of integrated management in the past years, and presents a formidable challenge to the authorities for the rehabilitation and restoration of the environment and the city.

One of the problems of the past has been the inability of the authorities to discourage construction at erosion prone locations. In Constanta, for example, there is a beautiful casino which was built at the beginning of this century. Today, it protrudes into the sea like a small cape, and needs to be protected against the forces of nature by ugly rubble and concrete blocks. Every time the building is hit by an exceptionally severe storm, the fortification around it needs to be strengthened by ever larger blocks, and this fight between men and nature will continue until the building is irreparably damaged and men lose the battle. At Batumi a project is underway to fill the sea north of the port to acquire space for a new railway station, a project which will no doubt start a new war between men and nature and which is again due to insufficient input of scientific knowledge into the planning and decision-making process.

A marine mammal, the Mediterranean Monk Seal, one of the most endangered marine species, once abounded along the shores of the Black Sea. Today, the species has almost completely disappeared due to environmental degradation, pollution, habitat loss, starvation because of the decreasing amount of fish and the hostility of fishermen. What does this loss mean to a scientist, a clergyman, an environmentalist, a fisherman, a ship-owner or an industrialist? The value given to the conservation of nature differs significantly among various sectors of society and people have different levels at which they balance economic development and nature conservation. Consequently, when faced with the dilemma of choosing between economic development and nature conservation, the priorities of individuals varies considerably. This is another cause for conflicts in the coastal zone.

The collision of ships along the Bosphorus, one of the narrowest waterways of the world, has increased significantly since 1979 in parallel with increased maritime traffic and ship sizes. In 1995, nearly 1,500 tankers alone (approximately five per day) navigated through the Bosphorus. It is no doubt that this is a significant menace to the city of Istanbul and to the health of the marine environment. In Novorossiysk a new oil terminal is being planned ready to ship the Caspian oil to the European market through the Bosphorus. Such decisions, which have significant transboundary implications not only for the Bosphorus and Istanbul, but also for the other states of the Black Sea, the Sea of Marmara as a whole and the Aegean Sea, need to be made through an integrated, regional management scheme.

Living resources, water quality, and nature conservation are other important issues which require significant management efforts at the regional scale. There is no way to solve the water quality problem in the Black Sea before doing so for the Danube, and this calls for the cooperation of the eleven non-Black Sea countries who share the catchment basin of the river.

In summary, the sources of conflicts in the coastal zone may be due to one, or more, of the following:

1. Scarcity
2. Incompatibility of uses
3. Priority differences
4. Insufficient information
5. Legitimacy
6. Fragmentation
7. Value evolution in time
8. Regionality.

The precious coastal areas and resources are finite and limited, and numerous consumptive or non-consumptive use options compete for them. The scale of the conflict grows in the case of mutually destructive uses of the coastal zone. The differences in the priorities of the local citizens against the national interest, and of different sectors of a society (for example, the private sector, scientific institutions, religious institutions, non-governmental organisations and concerned people) also result in conflicts for the use of a specific coastal space or resource. The lack of sufficient scientific information input into the planning and decision-making process in the coastal zone, which may be due to insufficient knowledge of the decision makers about the functioning of coastal systems and the general poor understanding of coastal processes, may lead to ecosystem damage, coastal erosion and flooding, in addition to adverse affects on the already existing uses. These issues then become the root of new coastal conflicts. Too little or too much legislation, and the presence and involvement of too many agencies and the lack of integration of their management efforts are also responsible for conflict.

There are two further important types of conflicts over the use of coastal areas. One of these is generational conflict, due to the evolution of societal values as a result of improved education, economical development and international contacts on the one hand, and to decreased opportunities and losses for future generations on the other. The second type, which is especially important for enclosed basins such as the Black

Sea, arises from variations of coastal management practices of the riparian countries and from lack of integration and harmonisation of their management efforts. At the regional level, international conflicts may be highly significant because of very important trans-boundary impacts, especially in the sectors of fisheries, pollution control, marine transportation, and nature conservation.

It is evident that high levels of management effort are needed at local, national, and regional levels for resolving coastal conflicts in enclosed seas. In order to achieve this, the following emerge as the way forward for effective coastal management:

- The optimisation of economical and social benefits derived from a scarce resource;
- A pro-active approach in management, and planning for non-destructive and compatible use patterns;
- The integration of the interests of different levels of the society (vertical integration), democratic nature and transparency of the process;
- A high level of scientific input into planning and decision-making processes;
- Clear and good legislation and effective enforcement;
- The means to bring various development and conservation agencies into a common planning and decision process (horizontal integration);
- Planning with long term objectives, including the options for the future generations;
- Effective regional cooperation for addressing activities with trans-boundary impacts.

The task is not easy. The present management efforts in many coastal areas around the world are normally shaped by our needs today. However, the needs of our children will be substantially different, and the values and priorities which contribute to a better life for future generations will not be the same as those we have now. It is therefore essential that we approach managing the coast from a long term perspective, and always incorporate into our management schemes the implications of our actions for future generations. If we fail to do so then we will no doubt cause inter-generational conflict as well.

Managing the Coast: Room for Everyone

Sylvie Goyet

A workshop is being organised in Tribias on the island of Villa in the Solomon Islands. Some 40 people, including chiefs and traditional leaders, have come from coastal settlements around the bay. With the help of trained local facilitators, they are discussing the issues of over-exploitation of the lagoon fish stocks, the increased number of sea turtles, the disappearance of the dugong population and the pressures from logging and mining companies to buy concessions from the communities and exploit the forest and land. In other words, they are planning for better management of the natural resources. Fishermen, traders and farmers from the villages are spelling out their concerns, needs and requirements. A simple natural resource management plan will be drawn and the traditional local council will see to it that it is followed and adhered to.

This activity is a classic preparatory exercise which, under another terminology, location and scale, would be one step in an integrated coastal zone management process. Integrated Coastal Zone Management (ICZM), though understood as a government-led process, can be initiated by local communities, particularly if these communities are the owners of the land and resources. It should be a process by the people and for the people - both present and future - an aspect that is often forgotten by experts when developing plans and mechanisms. ICZM processes have been initiated in various countries for some two decades, but it is evident that the needs of society and local communities have often been neglected in that process, as has the ecosystem.

The need for an integrated process of planning

It is now widely recognised that sectoral activities produce environmental impacts, which, when combined, result in greater degradation than the sum of each individual sector's impacts. They handicap each other and damage the environment. On the coastal zone, evidence of these negative environmental effects are seen in marine and fresh water pollution, air pollution, loss of marine resources, loss of natural land resources, land degradation and destruction of historic sites. Traditionally and still prevailing in most countries, policies to reduce coastal degradation have been based on a sectoral approach and have therefore failed, often only transferring problems among the resources and sectors and unable to take into account the overall impact of coastal developments on resources.

The basic concept of integrated coastal zone management is that the decision-making bodies should decide and act in a coordinated and integrated way and on the basis of complete information about the needs and requirements of the various parties concerned, thus minimising the costs and losses arising out of uncoordinated management, sectoral conflicts and magnification of impacts upon the resources.

Preparation involves the identification of sectoral problems and the analysis of present and future uses of coastal resources and their interaction, including present plans and policies. The process calls for the establishment of institutional mechanisms for integrated planning and inter-sectoral collaboration, often in the form of an inter-sectoral committee. Plans and policies are prepared as a framework for decisions regarding the management of:

- land resources for siting facilities for urban, commercial, industrial, mining, tourism and conservation land uses;
- coastal waters for recreation, aquaculture, transport and mining;
- living marine resources;
- assimilative capacity of coastal land, waters and air to achieve environmental quality objectives;
- coastal protection (against erosion for example).

Appropriate tools and instruments are then developed to manage the coastal zone in an equitable and effective way. Traditional tools include: legal instruments, land-use plans, enforcement techniques, environmental and economic valuation techniques, economic instruments and ad-hoc committees and permanent bodies, plus the establishment of the coordinating committee itself.

Initiating bodies are traditionally government agencies and ICZM is, in many cases, a government-sponsored and long term process. Pressures for ICZM are often exerted on central governments by international agencies (in the case of developing countries' initiatives) or result from the realisation that a good share of the coastal resources are viewed and perceived as public goods.

Further down the stages of the ICZM process, decision-making fora, are, in most cases, government agencies (the political and legislative sector, the bureaucratic and administrative sector and the legal and judicial sector); they can also be economic and market sector organisations. In some, rare, cases, the decision-making forum may be the traditional council system or people's associations.

There are good reasons why governments are in a prime position to take the lead in ICZM. In the Black Sea region, countries separately cannot effectively address those sea-wide problems such as transfers of pollutants and sediments, maritime safety and, to a lesser extent, tourist flows, but would benefit from harmonised and consolidated interventions. Moreover, the historic and traditional influence of policies and actions of the governments has been determinant on the evolution of the coastal zone, with policies on transport, fisheries, environment, agriculture, energy, industry and tourism, for example. Black Sea countries have also been fighting the same plights: beach erosion, decreased stocks of commercial fish and sanitation problems.

Only when issues directly involve local resources is the process more commonly taken on by local groups who are most willing to act when they perceive themselves to be direct stakeholders, as they do in subsistence economies. A key to involving local communities and people in ICZM planning and implementation would seem to be about enhancing their awareness of the benefits of cooperation affecting their direct or indirect use of the resources.

For increased equity among stakeholders

The key instrument in an ICZM approach is the setting up and functioning of an inter-sectoral committee or coordinating body entrusted with the identification of stakeholders, the study of their needs and requirements and the weighing of stakes and responsibilities. It is critical that all key stakeholders are represented in this committee, including those delegates standing for future generations and silent groups. Inter-sectoral committees are often established and staffed by those same bodies that have interest in the process: bureaucratic and administrative units and private sector conglomerates. As a result, less vocal and influential parties are often 'forgotten'. Delegates could include members of non-governmental or citizen-based organisations, professionals from social studies and environmental management and representatives from institutions such as the Church.

Under-representation of future generations

A proper ICZM process would include baseline studies and analysis of the characteristics of the ecosystem, including its carrying capacity, past and future likely trends, indicators of ecological health and threats. Such studies should also define minimum ecological thresholds below which the ecosystem as a whole or part of it would no longer function effectively and would be unable to regenerate to healthy standards.

To assess the impacts of development projects, programmes and policies, Environmental Impact Assessments (EIA) must be systematically carried out and efforts deployed to guarantee implementation and enforcement of EIA legislation. Minimum ecological thresholds must be set and monitored regularly and an 'alarm system' installed to ensure that prompt remedial action is taken to safeguard the affected environment.

Under-representation of local inhabitants

Balancing and weighing up the different interests is often not equitably observed, the losers usually being the local communities. Surveillance of the social integrity of the society must be introduced into the ICZM process in a way that will contribute, in the long term, to reinforcing the cohesion of the community, to enhancing their ownership of and responsibility for the resources and to strengthening their confidence in the process itself and their cooperation in its implementation. The incorporation of public awareness activities early on and harnessing inputs and participation from a larger number of stakeholders can contribute to this process.

The impacts of the proposed development and policy changes upon the social systems must be monitored to help determine the levels of disturbance the system can endure and the minimum requirements that would have to be maintained to avoid abrupt disruption in the structure of the society. Indicator values would include crime rate, unemployment level, family structure, growth rate and nutrition patterns.

Examples of neglected social integrity have shown that overlooking it may lead to resentment and non-cooperation and ultimately to violent rejections in some cases. The local population would be likely to show little cooperation when public access to space and resources is not secured (public beach access for example), historical and cultural grounds are not respected (taboo areas or religious places) and valuable sites, buildings and monuments are not preserved.

To ensure the integrity of the social systems, three measures are therefore recommended:

1. Baseline/groundmapping social studies be carried out and include identification and definition of social threshold values;
2. Social indicators be measured or assessed periodically and a watchdog system be put in place to monitor the evolution of these indicators (NGOs and community-based organisations can take a lead role in this exercise);
3. Public awareness activities be scheduled throughout the ICZM process from preparatory to enforcement and evaluation stages.

Recommendations

To further an equity in the use of the coastal resources and integrity of the social, economic and ecological systems, it is recommended that :

- Baseline/groundmapping studies be systematically carried out and ecological minimum thresholds be proposed as indicators of the good health of the environmental and social minimum thresholds as indicators of stability and development of the society. Together with criteria for optimum natural resource use and economic efficiency, these would guarantee the social and environmental integrity of the coastal zone as well as the sustainable development of the area;
- True representativity of all stakeholders be encouraged. That implies a greater responsibility and role for community outreach institutions and organisations;

- Public awareness activities be scheduled and incorporated in the process;
- A monitoring mechanism be put in place to assess the effectiveness as well as the impacts of the process and plans upon the ecological, social and economic systems;
- Tools and instruments for ICZM be developed across the three areas of interventions: at the institutional level (for example, legal instruments, LUP, enforcement techniques, conflicts resolution), at the economic level (for example, environmental valuation, economic instruments) and at the ethical level (for example, environmental education, public awareness activities, general rights for information, participation and standing rights), so that adequate signals, in the form of accurate and complete information, are sent to the decision makers.

The Black Sea Environmental Programme has endeavoured to facilitate a partnership approach in environmental decision-making fora, in particular as it refers to ICZM. It has encouraged the participation of all levels of the formal government system and sectoral agencies and has also helped strengthen the capabilities and confidence of non-governmental organisations to take an active part in the process. In principle, ICZM can secure the integrity of coastal communities and guarantee equity among the stakeholders. Giving those stakeholders a proper voice and ensuring that they use it is vital to its success.

In the light of current debate, there is good argument to now invite the Church to develop practical actions to follow-up on the Assisi Declarations, drawn up in 1985 by five prominent religious institutions to guide conservation-oriented thinking and activities in their respective faiths. Such actions could include raising the public's awareness of environmental issues and threats, promoting the responsibility of individuals for environmental care, and contributing to the setting up of a framework and a process for guiding environmental management and ICZM in particular.

Bibliography

Assisi Declarations, WWF (1985).

The Baltic Sea Joint Comprehensive Environmental Action Programme.

BSEP (1996) Coastal Zone Management - Integrated Policies, Summary report on BS-ICZM prepared by ICZM IAC, Krasnodar, Russian Federation, OECD.

Ercmann, S. (1996) Study on the feasibility of drafting a European Integrated Coastal Management Instrument, paper presented at the Group of Specialists on the Protection of the Coasts, Strasbourg, 6-7 June 1996 and at the First Inter-Parliamentary Conference on the Environmental Protection of the Black Sea, Istanbul 10-12 July, Directorate of the Environment and Local Authorities of the Council of Europe.

European Commission (1996) Demonstration Programme on Integrated Management of Coastal Zones, February.

Rhode Island Workshop (1995) Educating Coastal Managers, March 4-10.

UNEP (1995) Guidelines for Integrated Management of Coastal and Marine Areas, UNEP Regional Seas Reports and Studies no.161.

Messages from the Black Sea

Trabzon to Thessaloniki

I LEARNED as a child that I should do unto others as I would have them do unto me. 'Others', I was told, include birds, turtles, fish, trees and all other living things. Human beings tend to take up a lot of space and do not photosynthesise. We must eat and inevitably must displace other forms of life to build our houses and communities and great cities, but to do these things blindly, without considering the consequences to the natural systems that support us, is to court ethical, ecological, and, eventually, economic bankruptcy.

The Black Sea is bordered by six countries and during this trip, we touched - and were touched - by them all. Each day distinguished delegates from the country at which the boat was docked joined the symposium, including leading politicians and leaders of the Orthodox Church. The visit of the Ecumenical Patriarch was a major event, often celebrated by a warm welcome from people and clergy on the dockside and later by church services. These were the first ever visits by an Ecumenical Patriarch to the main cities of the Black Sea. - *Dr. Sylvia Earle*

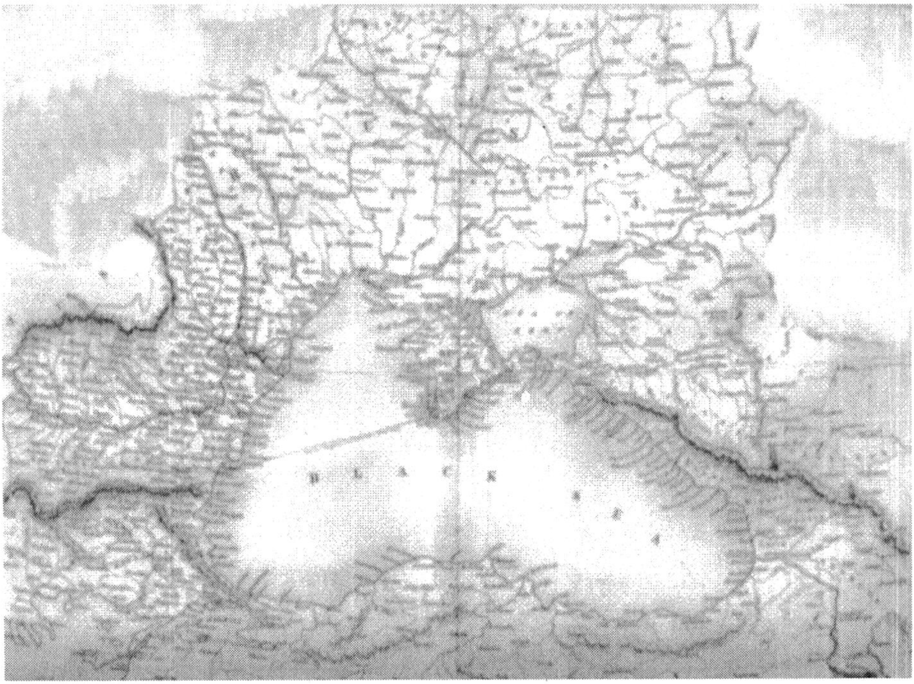

Trabzon, Turkey

This ancient port-city, once Trebizond, was for many centuries the economic and political capital of the Black Sea's southern coast. It was a meeting point for many cultures and trade routes. Originally a Greek colony (Trapezos), the city became the centre of a full-scale Greek territorial settlement, extending up to the watershed of the Pontic Alps, which was to last from the sixth century BC until the expulsion of the Christian population of the Pontos in the 1920s. Between 1204 and 1461, Trebizond was the capital of a pocket empire in the Byzantine period, when the Grand Commenian Empire enjoyed fame and prosperity as the terminus of a 'Silk Route' coming up from Persia and connecting Europe with central Asia and China. A second time of prosperity began in 1829 when treaties reopened the Black Sea for commerce and the city became a hub of world trade until the Suez Canal was completed in 1869. The modern city of Trabzon is an important fishing port and trading centre which has entered into a period of rapid growth, fuelled by the recent opening of important trade routes to Russia and Georgia.

THE PHENOMENON of climate change likewise respects no borders. The planet's seas and forests are part of the global commons, not just any single nation's prized commodities. The United Nations is working with the countries of the Black Sea region to protect the marine and coastal environment and to promote its management and development. The Black Sea is covered by the United Nations Environment Programme's Regional Seas Programme and other United Nations agencies are providing their expertise as well. Let us continue this partnership for the sake of all those whose lives and livelihoods, cultures and spirits are deeply entwined with this renowned body of water. -*His Excellency Mr. Kofi Annan, Secretary General of the United Nations*

I BELIEVE this symposium will accelerate the efforts aimed at rendering the Black Sea a fertile area, generously offering its rich resources to the people living around it. Future generations should find a Black Sea with better characteristics. -*President of Turkey, Mr Suleyman Demirel*

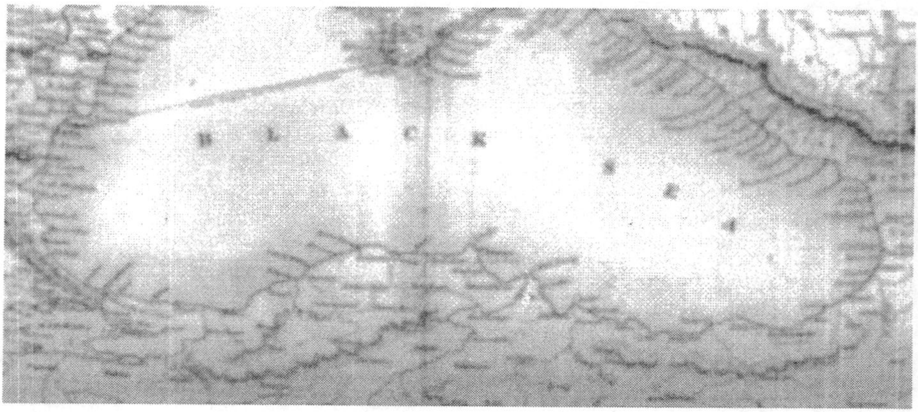

Batumi, Georgia

Situated in the independent Republic of Georgia at the south-eastern corner of the Black Sea, the port of Batumi is also the capital of the autonomous region of Adjaria. Its climate is warm and notoriously rainy. Batumi is gravely dilapidated and its problems include the presence of thousands of indigent refugee Mingrelian and Georgian families who fled from Abkhazia in the early 1990s after the Abkhazian war of independence against Georgia. The city also suffered during the civil war following Georgian independence. Its infrastructure has been partly wrecked, health standards are falling and there is serious coastal pollution. In short, Batumi is a good example of all the problems of a society in transition from Communism to a market economy. It is also a good example of how religious communities can live together, for there are well-established communities of Christians, Jews and Muslims living in harmony with each other in Batumi.

THIS IS a real historical event - science and religion hand in hand starting to settle the main task of mankind and secure the preservation of life on the planet. Both sides are reaching the conclusion that by peaceful existence, co-operation and friendship, we can improve the life of the present generation, which can serve as a base for the healthy and peaceful existence of future generations. - *Mr. Aslan Abashidze, Head of the Supreme Council of the Adjarian Autonomous Republic*

CURRENTLY GEORGIA is elaborating the National Black Sea Action Plan and integrated management of the coastal zones. The successful completion of these programmes will depend on the solution of the country's political, economic and social problems. The region is disturbed not only by economic fragility but also by ethnic, social and political conflicts. One of the principles of sustainable development is the prevention of armed conflicts. The time has come when the problems of environmental protection and principles of sustainable development should be raised at the negotiating table, for these issues are facing all of us. - *Ms. Nino Chkhobadze, Minister of the Environment for Georgia*

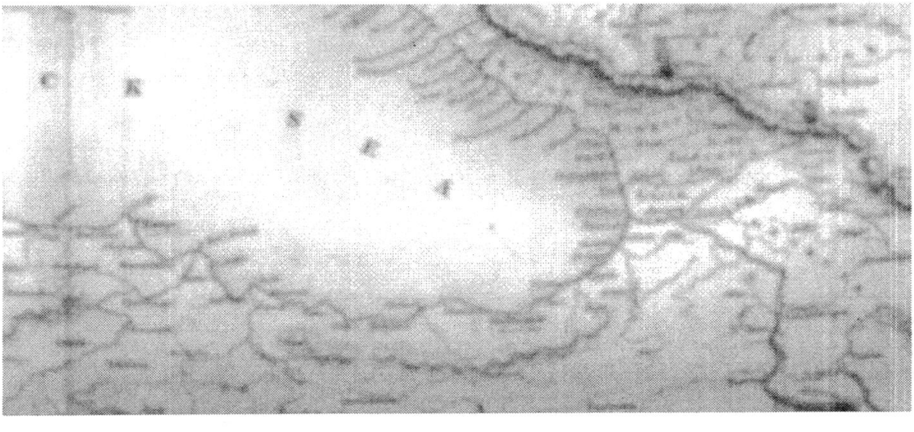

Novorossiysk, Russia

Since the collapse of the Soviet Union, this is now Russia's only major harbour on the Black Sea. The city was founded by Catherine the Great at the end of the eighteenth century, on an inlet of deep water on the Kuban coast. It soon became a large commercial port and naval base. In 1920, at the end of the Russian civil war, the British and French navies evacuated the Cossack and White armies from Novorossiysk. In the Second World War it was the site of savage fighting against the German invaders. Novorossiysk today is a handsome nineteenth century port city, with broad avenues, disfigured by the gigantic cement works which belches smoke and has over 70 years stripped away the limestone mountainside above the town. Today, it has a new prosperity as the sea terminus for one of the pipelines from the Caspian oil fields; the project for a new offshore oil terminal a few miles to the west has roused local anger and led to the formation of a lively ecological protest movement. Beaches in the area are heavily polluted by oil spills and by the dumping of domestic and hospital waste.

NOVOROSSIYSK is the main strategic point for the transhipment of Russian trade in the Black Sea. According to the Federal Development Plan for Krasnodar, Novorossiysk is to be developed as a trade and transhipment centre of international standard. During the development phase, special attention must be paid to the fact that there should be a balance between the economic and social infrastructures. Planning must take into account the historical, national, geographical and climatic conditions of the area in order to overcome the existing crisis. There are already current projects which focus on stabilising the situation in the fields of health, culture, education, ecology and religion. - *Mr. Valery Prokhorenko, Mayor of Novorossiysk*

SUPER-TERRORISM and pathological greed are the motivating forces of a few of our fellow men planning the construction of an oil terminal twice the size of the existing one in the city, as well as a pipeline that will go through the Caucasian mountain forests and the tectonic slopes of the mountains, onto the sea bottom, where earthquakes are measured more than ten times a year. The pipeline will also pass through children's health resorts and through lands where the mortality rate of the inhabitants is seven times greater than the average indicator for Russia. Our civilisation is far from being number one; gold is not edible nor can it save anybody from ecological catastrophe. After the last two accidents in our city, the plankton, geese and sea gulls died in large quantities. That is our future - sea without seagulls and the trees in the city diseased. - *The Association for Saving the Black Sea, Novorossiysk*

UPON COMPLETION, our pipeline will provide a critical outlet for Russian and Kazak crude oil to world markets which in turn will create enormous revenues for the region. Over the 40-year lifespan of the pipeline, Russian and Kazak tax and dividend revenues will exceed 31 billion dollars, providing much needed jobs, infrastructure and social benefits to local communities. The pipeline will meet or exceed the highest international standards for construction methods and environmental protection. We believe that local communities must be involved in this project at every level. CPC has established a 24-hour information hotline and is conducting regular public meetings throughout the regions along the pipeline route to address the concerns of local government authorities, businessmen, journalists and ordinary citizens. - *Mr. V.S. Stanev, General Director, Caspian Pipeline Consortium*

Yalta and Odessa, Ukraine

At the foot of the majestic Crimean coastal mountains, Yalta is the oldest and most famous seaside resort for Russians and Ukrainians. Remnants of the Tatar population, deported to Central Asia by Stalin, have returned to the area, but the majority of the population of Yalta is Russian. (Crimea, which is now semi-autonomous within Ukraine, has been the focus of dangerous confrontations in recent years between Russia, still resentful at its loss, and Ukraine.) Almost all the great Russian writers spent time at Yalta, the location of many of Chekhov's stories, although the town's economy is now suffering from overcrowding and lack of investment due to Ukraine's chronic crisis of transition. Palaces near Yalta were used for the 1945 Yalta Conference, which settled the fate of East-Central Europe for half a century, and in 1991 the Soviet reformer and leader Mikhail Gorbachev was held hostage in his nearby villa at Foros during an unsuccessful coup d'état.

Odessa is the Marseilles of the Black Sea. Founded in 1794, this handsome and exuberant city became the main port through which wheat from Ukraine was exported in the nineteenth century to western Europe. It was always a multicultural city: its first governor and planner was the Duc de Richelieu, its first commercial language was Italian and its shipping belonged mostly to Greeks. Odessa also attracted a large and vigorous Jewish population (commemorated by the writer Isaac Babel). Today, Odessa is rapidly recovering from the disastrous years after the fall of the Soviet Union, and derelict buildings are being repaired and redecorated. The city is also the home of several marine biological and ecological research institutes, each with a small fleet of specialised vessels, which have been almost paralysed by the cancellation of their grants and the shortage of fuel for expeditions. A lively ecological movement sprang up in the last few years of Soviet power, mostly in order to resist the project for a new oil terminal. Odessa suffers from severe damage to marine life in Odessa Bay, the result of sewage discharges and eutrophication through fertiliser runoffs, and the water supply from the River Dniester is badly contaminated.

THE HUMAN is on Earth not as a stranger who came to receive a monetary profit, but as a careful owner who cultivates the Earth for future generations and takes care not only of his own profit, but also of the good of his neighbours and those far off. Moreover, the care of protecting the Creation of God in all its beauty and harmony is not only our practical task but also a spiritual and religious duty, a fulfilment of the commandment of God and a trial of moral feeling. The Black Sea region has suffered from many sad consequences of an unreasonable, selfish use of nature and this has been especially dramatic in our century. Today we must understand the need to work together for the transfiguration of this wonderful piece of land, for the improvement of the condition of the Black Sea, the pearl of our planet. - *Alexey II, Patriarch of Moscow and All Russia*

EVEN THOUGH the Republic of Moldova has no direct access to the Black Sea, the Rivers Nistru and Prut that form our border, as well as other waterways that flow into the Danube, give us a special responsibility for the ecological state of the region. Inter-governmental agreements have been concluded with the relevant bodies of Romania and the Ukraine. We have made arrangements to participate in the study and implementation of a tripartite Moldovan-Romanian-Ukrainian agreement concerning the reasonable use and protection of the River Prut, and to give a new impetus to the Moldovan-Ukrainian ecological cooperative programme for the River Nistru. We are also carefully studying the proposals for the creation of a national park. Unfortunately, the thorough exploitation of these possibilities is hampered by a lack of economic means. We know that other states of the Black Sea region are also facing the same problem. We should therefore study together how we can achieve the gradual elimination from the Black Sea region of all kinds of weapons of mass destruction, and how we can ensure that the sums of money thus saved are used for ecological development. - *Mr. Petru Lucinschi, President of the Republic of Moldova*

I BELIEVE that your work will not only have a positive effect towards the solution of specific problems associated with the protection of the environment, but make a significant contribution to the development of the relationship between the peoples of the area. - *Mr. Valeri Poustovoitenko, Prime Minister of Ukraine*

A GREAT PART of the pollution in the north-west area of the Black Sea is caused by the substances and materials transferred from the great European rivers, mainly the Danube, which discharge on the coast of the Ukraine. The sea water on the coast of the Ukraine is heavily polluted because the discharge and treatment systems for home and industrial wastes of the 21 coastal cities and 35 towns of the Ukraine are either very old or non-existent. Serious ecological damage is also caused by the seven shipyards and 20 transhipment points for trade goods, petrol and oil products. Pollution of the air is caused by the coastal industries, the shipyards and transhipment points. The effect of pollution on the coast of Ukraine becomes obvious in the recreational and resort zones with epidemics and outbreaks which have forced the Ministry of the Environment of the Ukraine to close the resorts. Three international centres have now been established in Odessa to gather and evaluate information related to the status of the Black Sea environment and establish emergency response units. - *Mr. P. Bodelan, Chairman of the Odessa Region and Member of the Ukraine Parliament*

Constanta, Romania

Built on the site of the old Graeco-Roman colony of Tomi in which Ovid passed his exile, Constanta is a large and well-designed city with Romania's main commercial harbour. An hour's journey to the north lies the Romanian section of the Danube Delta, a huge labyrinth of reed beds, side-channels and small islands, some of which have human settlements. The Delta, which is not only the river's natural filter before it reaches the Black Sea but also one of Europe's richest nature reserves, was badly damaged in the later years of the dictator Ceausescu's rule, when he decreed that it should be drained and transformed into agricultural land. Some parts of the Delta, lost in this way, are being restored now, but the region also suffers from pollution - industrial waste, excess fertiliser nutrients, crude oil waste coming down the river from as far upstream as Austria and Germany, and from the effect of the big dams at the Iron Gates and Gabcikovo which regularise the flow and thus reduce the Delta's natural seasonal rise and fall.

THE BLACK SEA was and is a bridge between all those who during their millenary history have lived together and have created civilisations which have survived across time in the conscience of many generations. It would be absurd to suggest that we should stop the progress of science. What we can and should do is to use science more efficiently to ecological ends. If we decide to do this we can control and block the pollution of nature and we can learn that it is the responsibility of each citizen to live in the spirit of ecology and to respect above all the right of his/her fellow human beings to live a clean life in its fullness. Romania is and will always be in the first line of this struggle. - *Mr. Emil Constantinescu, President of Romania*

IT IS EQUALLY important that besides the administrative, legal or technical measures to which one appeals for improving the ecological environment of the Black Sea, a special emphasis be put on the spiritual, religious aspect of this salutary action. Thus we reckon that one will be able to meet these ecological concerns from a religious point of view if one takes into account the words of Genesis which witness the spirit of God in Creation. It is in this sense that the Romanian people emerged in history as a Christian people, understanding God as a Sun that sends out light, life and love, the uncreated divine energies, over the whole of creation. Both God's transcendence beyond creation and his immanence in creation are very important for the efforts we make for preserving ecology. - *His Beatitude Teoctist, Patriarch of Romania*

THE BIOSPHERE park of the Danube Delta has evolved in one of the least natural regions of the world and is subject to a triple regime of the RAMSUR Agreement, the UNESCO agreement and the international network of natural biosphere parks. The convergence of three golden ecological principles is needed: that should be saved which can be saved; that should be restored which has been destroyed; and possible disasters should be anticipated and averted. - *Dr. Janos Rolond Kiss, General Secretary of the Department of the Protection of Rivers of the Ministry of Rivers, Forests and the Protection of the Environment of Romania*

Varna, Bulgaria

Varna is a medium-sized port with a shipyard, a naval base and a magnificent park above the harbour, filled with flowering bushes and enhanced with antique torpedo boats from the pre-1924 era. Transition to a free-market economy seemed to be making only slow progress; the strolling crowds in Varna's wide streets were leisurely and friendly, but the buildings and pavements were shabby and cracked. This is one of those Black Sea cities whose life may soon be transformed by the impact of the Caspian oil boom, if - as seems likely - it becomes a base for service vessels and oil-related industries during the construction phase.

THE BULGARIAN community, whether represented by its state authorities, its industrial circles or its religious and cultural organisations, is conscious of the share of responsibility it has for the reduction and ultimate elimination of effluents that harm the Black Sea. It is with satisfaction that we have been following the activities of the Ministry for the Environment in our country, which is making efforts to study and limit the effect of these factors which are causing damage to the marine environment. In our community the harmful exploitation of nature, which is the creation of God, is no longer tolerated. It is incumbent on us to use the material world which God has entrusted to us in a beneficial way, not exploit it mercilessly. We should hand it on to the generations that come after us, not as a wasteland but enhanced and with a greater capacity for supporting life. - *Patriarch Maxim of All Bulgaria*

OUR COUNTRY is aware of the urgent need for integrated management of our coastal zone. We are the first Black Sea country with a specific regulation in force and with an administrative structure for coastal management already in place. A draft law on integrated coastal zone management has been prepared. A National Action Plan is being drafted. A geographic information system is being created and the coastal municipalities have new regional development plans. These results have been achieved with the financial support of the World Bank and the assistance of US specialists. The Danish Government supports the modernisation of the Varna wastewater treatment plant. Varna has the best laboratory in the Black Sea region for analysis and identification of oil hydrocarbon in marine waters, sediments, biota. - *Ms. Evdokia Maneva, Minister of Environment of Bulgaria*

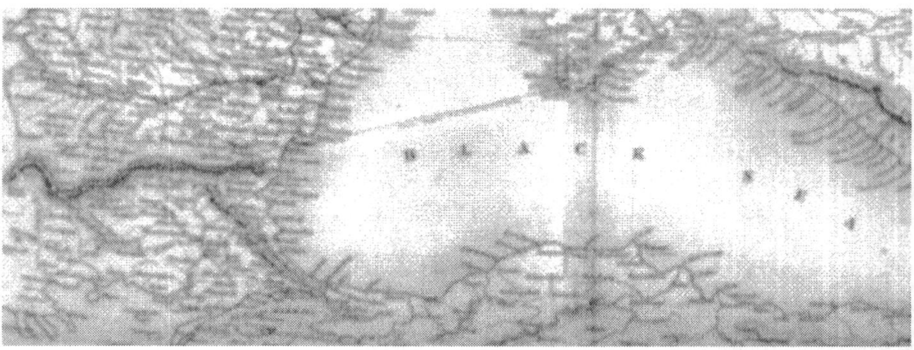

Istanbul, Turkey

The great city is facing multiple economic and ecological crises and choices. The marine life in the Bosphorus is in headlong decline and many fish and invertebrate species are in danger of extinction. The main cause is overwhelming pollution, made worse by local over-fishing, which is destroying the habitat and breeding-grounds of fish in the Sea of Marmara. Some of this contamination comes from shipping passing through the Bosphorus, some from on-shore industries or from the discharge of untreated sewage into the Golden Horn. While the city itself is expanding at uncontrollable speed - there is no exact number for its population - the greatest single threat to Istanbul today comes from the rising congestion of shipping passing up and down the Bosphorus, particularly as it is now proposed that supertankers serving the Caspian oil fields should use the Narrows as their route to the Mediterranean.

ACCORDING TO our sublime religion, and the teaching of all monotheistic religions, everything that exists in the world has been loaned to us by God in His great goodness. Nature in its proper state has an order, a harmony, a measure, which is rightly called ecological balance. The ecological order which God has given us is constantly disturbed by the uncontrolled and thoughtless consumption of natural products and primary materials. It is a historical, religious and moral necessity for us to take all necessary measures to put a stop to this evil tendency. Although technological and economic development must continue, it is essential that the natural environment and biodiversity be preserved. Religious faith includes basic precepts and examples relating to the theme of the environment. We must certainly profit from these commandments, accepting as the divine will that the whole of the universe is merely on loan to us. This is a conviction that we should disseminate among people, especially the young, to whom we entrust our future. - *His Excellency Mehmet Nuri Yilmaz, Head of the Department of Religious Affairs*

TURKEY HAS made a contribution to the Black Sea Environmental Programme from the outset by acting as host and offering practical support. We provided 40% of the expenses of the Secretariat of the Black Sea Commission for the first three-year term in fulfilment of our commitment. Neither the efforts of a single country nor common efforts of the six coastal nations can suffice in the protection of such an important resource. I therefore invite the administrators of those nations which are situated within the Black Sea basin, in addition to coastal nations who have been dumping any kind of waste into the Black Sea, to cooperate and give their respective support to our ongoing efforts. At present about 30 million tons of crude oil is being transported every year through our straits. Transportation of central Asian oil to the Mediterranean will transform this area into an oil lane. Such an addition to a transport channel which is already threatening life, property and environmental security should never be accepted by any environmentalists and cannot be reconciled with the principle of sustainability. - *Her Excellency Dr. Imrem Aykut, Minister of Environment, Republic of Turkey*

Thessaloniki, Greece

This is the ancient capital of northern Greece, under Ottoman control from 1430 to 1913 and now the nation's second city after Athens. As the vessel entered the port, after passing close under the monasteries of Mount Athos to the south-east, the ships in the harbour greeted the Ecumenical Patriarch and dense masses of people could be seen waiting along the waterfront. Thessaloniki was enjoying the title of Cultural Capital of Europe in 1997. There had already been many months of celebrations and cultural events and the city was in high festival spirits when the symposium docked for this last port of call.

GREECE IS an active and effective member of the Black Sea Cooperative Programme. The city of Thessaloniki is the seat of the Bank of the Black Sea Cooperative Programme and the permanent seat of the Secretariat of the Black Sea Cultural Network of Cities and Arts. Sensitivity towards the environment is above all an expression of the solidarity we feel with past and future generations. The decisions that are made are not confined to the chronological horizons of the current generation or of the immediately succeeding generations, but are made - or, at any rate, should be made - within a much greater perspective. The crucial issue is the triangular relationship between humanity, nature and time. This is both a theological and a historical relationship, but above all it is a political relationship, because the relationship between humanity and nature invites us to make choices which involve the weighing of values, choices between the values on the one hand represented by progress and development, and those on the other connected with the quality of human life. - *Mr. Evangelos Venizelos, Minister of Culture of Greece*

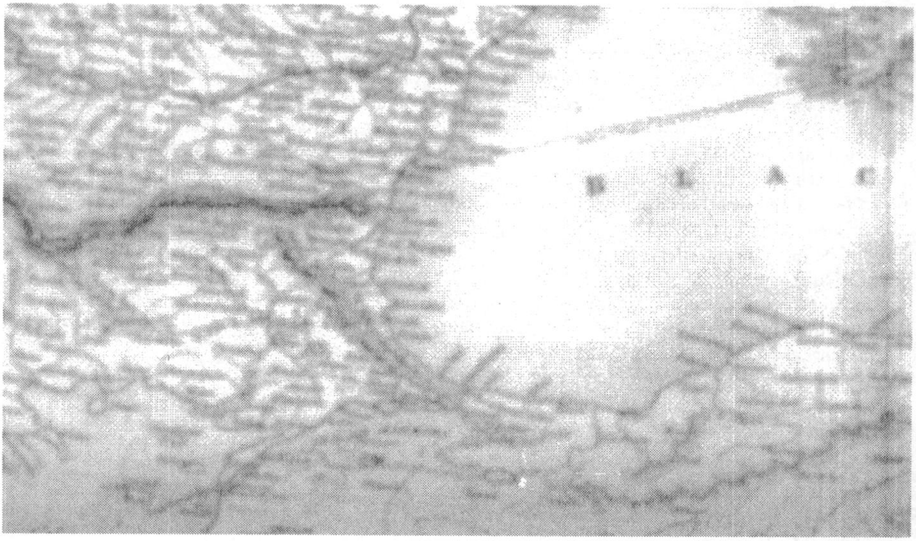

Chapter 5

The Human Perspective

Introduction

Richard J.C. Chartres

In our travels in the Black Sea region, participants in the symposium encountered again and again the puzzle that although there might be theoretical understanding and incontrovertible scientific evidence to pinpoint the causes of some serious environmental problem, this did not by itself translate into effective action. Often we know what ought to be done, but we do not do it. The papers in this section deal with different kinds of human perception and the translation of those perceptions into ecologically beneficient action.

The journal *Nature* has recorded the results of a fascinating experiment in which kittens were reared in an environment dominated by horizontals. After prolonged exposure to this environment, they were released as adult cats into a normal domestic setting. Thereafter they experienced considerable difficulty in perceiving verticals and were forever cannoning into chair legs and other pieces of furniture. The truth seems to be that although the capacity to perceive the vertical is present in the brain, if this faculty is not triggered and developed in early life, then it can very easily atrophy.

Something of the same seems to have happened with the capacity for spiritual perception in the experience of many modern people. This experience of the spiritual is very frequently in the religious classics of various world religions likened to an ascent or an encounter with the vertical dimension of life. Ravi Ravindra paints a refreshing picture of the progress to spiritual enlightenment in his paper, 'Healing the Soul'. His emphasis is on personal, spiritual perception and this is complemented by the social vision which informs the paper by Bishop Morelli which follows. He makes a passionate plea for a deepened awareness of our social solidarity with the poor who have to carry so many burdens of pollution and waste. We live in a world in which 20% of all human beings consume close to 75% of everything that is produced. Bishop Morelli's contribution to the symposium demonstrated the energy that can flow when things are seen from the perspective of the 80% of humanity who are left with a bare quarter of all that is produced.

Severn Cullis-Suzuki's piece sees environmental damage from the perspective of intergenerational ethics. It is clearly unjust that our children should be presented with the bill for the way we live now. With delightful humility, Severn remarks that she is 'now 17 and over the hill', but she is still able to give a voice to the generation that will soon inherit our world.

Humility, in its profoundest sense, is down-to-earth humus earthiness and this kind of humility is also a feature of the papers by Zaki Badawi and Naftali Rothenberg. Dr Badawi, in recommending a precautionary approach in the face of the power of modern technology to make profound changes in the environment, draws attention to the tradition in Shariah law that 'the wastrel and extravagant can be deprived of freedom of action under the law and placed under the control of a guardian'. He underlines the limits within which we are set to work in this world but his contribution is also a plea for humility and self-criticism on the part of religious authorities.

Rabbi Rothenberg outlines a religious 'third way' between unrestricted human domination of nature and those who tend to worship nature and relegate human beings as merely one element in nature and different only in the degree of their destructiveness. The religious 'third way' is expressed in the placing of responsibility on humankind to find the balance between 'tilling' the world, that is, intervening in nature in order to help it flourish, and 'keeping' it, which means the avoidance of needless destruction, particularly of the kind that cannot be undone.

Reflection on these themes in a political context is offered by William Reilly, who concludes that arguments based on purely rational and scientific calculations rarely stir people to action. He quotes Jacques Cousteau, the eminent pioneer of underwater studies, who once said 'in the end, people protect only that which they truly love.' It is a pithy summary of the truth which reverberated throughout the whole symposium.

Healing the Soul

Ravi Ravindra

Om! Asato ma sad gamay
Tamaso ma jyotir gamay
Mrtyor mamrtam gamay.

Om! Lead me from untruth to Truth
Lead me from darkness to Light
And from death to Eternal Life.

(*Brihadaranyaka Upanisad* I.3.28)

Healing comes from most high

The words for *truth* and *reality* are the same in Sanskrit, as in many ancient languages. In the invocation above, *sat* is both truth and reality. Only the real is true, and truth alone is real. When we live in fantasy, imagining this or that, driven by fear or self-importance, we fall into darkness and fragmentation. Truth alone can bring us to light and make us whole.

The soul is oriented towards the Vastness (Brahman) much as a magnetic compass always seeks the north pole. However, impediments, compulsions and obstructing forces can interfere and one can be alienated from the Highest. Integration requires a spiritual discipline. Connecting with the Highest is the aim of all religious and spiritual paths. The great traditions, in wondrously different ways, have maintained that the Highest Reality - variously labelled 'God', 'First Principle', Original Mind' or simply 'That' - is Truth and is Love.

This insight needs to be continually regained, lived and celebrated. Only when and wherever this realisation is made concrete, is there an abundant life of the Spirit. Spiritual disciplines are all concerned with integration and wholeness; above all with the integration of truth and love. Love is required to know truth, and knowledge of truth is expressed by love. Whenever truth and love are separated from each other, the result is either sentimentality or dry intellectualism in which power is divorced from compassion. Partiality always carries seeds of violence and fear in it. Thus in the name of 'our loving God' many people have been killed, and many destructive weapons have been developed by a commitment to 'pure knowledge'.

One can see different emphases in the great traditions. Those oriented to insight and truth, such as the Hindu-Buddhist spiritual traditions, have elaborate descriptions of various levels and states of *consciousness* whereas in the faith and love-oriented Biblical tradition, the emphasis is much more on the various levels of *conscience*.

This major difference between the Biblical and Indic traditions is related to another major difference. In the Biblical tradition the root cause of the human predicament is considered to be an assertion of human *self-will* as opposed to the will of God as revealed in His Commandments. In the Indic traditions, on the other hand, the root cause of the human difficulty is regarded as ignorance, which in its turn gives rise to suffering (*dukkha*) and to illusion (*maya*). In the Biblical tradition, submission of our will in obedience to the will of God is called for. In the Indic tradition, the requirement is for the sword of gnosis (*jñana*) which will cut the knot of ignorance.

Another fundamental difference follows from this. In the Indic traditions, to hold on to a separate individuality is regarded as a mark of ignorance, whereas in the Biblical tradition a lack of individuality - even in the presence of God - is seen to be a mark of lack of responsibility. In one case, the traditional emphasis is very much on the *oneness* of all there is, whereas in the other case the emphasis is on the *uniqueness* of human beings. Both *oneness* and *uniqueness* are derived from the same root, but their meanings diverge radically. The traditions which hold the ideal of oneness are insight-oriented and have developed a great deal of wisdom about various levels of consciousness, which have to do with degrees of steadiness of attention and gradations of clarity of perception. The traditions extolling uniqueness are oriented towards faith and obedience and have a great deal to say about individual responsibility and moral conscience.

In fact there is an intimate connection between conscience and consciousness. The awakening of conscience is the preparation for an enhancement of consciousness. It is not possible to come to a higher state of consciousness without coming to a higher state of conscience; those who are in touch with higher levels of consciousness naturally manifest largeness of heart.

Spirit, soul and body

Upon returning from a battle, Prophet Mohammed once said to his disciples, 'We have returned from the little holy war to the great holy war'. When the disciples pressed him for clarification, he said: 'The little holy war is a war against the infidels, but the great holy war is the war against one's own soul.'

In modern Christian understanding, the soul is considered to be so wholly good and spiritual that the idea of waging a war against it is likely to sound like siding with the devil. It is therefore useful to remember that in many ancient traditions, and certainly in early Christianity, Greek philosophy and Indian thought, a developed or mature human being was regarded as having three distinct parts. These were Spirit, soul and body, or to use St. Paul's terminology *Pneuma*, *psyche* and *soma*. *Pneuma* (Spirit) is suprapersonal, manifesting itself in or through a person or in other ways, whereas *psyche* (soul) belongs to the person. Thus it makes sense to speak of 'my soul' but not of 'my Spirit' just as one cannot speak exclusively of 'my God'. On the other hand, one can speak of 'the Spirit' or of the 'Spirit in me'. No one can appropriate the Spirit in any possessive or exclusive manner. From a spiritual point of view, the best one can hope for is to be appropriated by the Spirit so that all of one's energies - mental, emotional and physical - can serve the will of the Spirit.

Over the centuries this three-fold division gradually shrank into a two-fold one: spirit and nature, mind and matter, soul and body, or higher nature and lower nature. Since the time of Descartes, 'soul' has been regarded in the western world as essentially spiritual rather than natural. A similar process took place in India. There, psyche is considered to lie in the realm of matter, and is therefore subject to the laws of material nature and amenable to natural scientific inquiry. The psychic functions, such as thoughts and feelings, and the psychic phenomena, including the so-called paranormal ones, are in the realm of *Prakriti* (Nature), the domain of materiality and causality. According to Indian thought, so-called miracles, those mentioned in the Bible for example, are natural, rather than supernatural or spiritual, even though they are unusual and extraordinary. It is good to remind ourselves here that what we know about nature is not all there is to nature.

Thus as Prophet Mohammed said, the war against the soul is a war against one's own lower nature. Similarly, when Jesus Christ says in Luke 14:26, that he who does not hate his psyche (which is to say his soul or self), cannot be a disciple of his, 'soul', as in Indian thought, belongs to the lower realm. However, it is within this lower realm (also spoken of as the realm of the 'world' or the 'flesh') that the possibility of a new life, a birth into the higher world of the Spirit, lies. In the three-fold division of a human being, a division which is philosophically clearer and more satisfying than the two-fold one, the struggle between the spiritual and carnal natures is waged in the psyche of each human being. A human being can side completely with one or the other force, becoming wholly an animal or an angel, or he can keep both of them in their proper balance and be truly human.

Healing the soul

The soul is not only the arena but also the faculty of awareness of the struggle between our carnal nature and our spiritual nature. This is manifested primarily in the ability for self-reflection, and in an aspiration to be oriented to the whole. Although one may wish to maintain that every human being has the potential for inner awareness or self-reflection, in practice not many of us have the luxury or the requisite freedom from fear and self-importance to engage in much inner awareness. Inner awareness can, however, be cultivated.

A great deal of emphasis has been placed by the different traditions on self-knowledge which transforms and heals the soul. Self-knowledge is a prerequisite to, if not synonymous with, knowledge of God. In fact, it appears that the only way to know God is by self-knowing, dying to one's superficial self and being born to a deeper self. There are differences in detail and emphasis, but there is no other point in which there is a greater unanimity of principle among the various masters of spiritual becoming.

Obviously, there are levels of self-knowledge which correspond to a development or deepening of being. Different traditions refer to these levels in many ways: as a seven-storied mountain, as rungs of a ladder, or as levels of consciousness. Within ourselves at the level of ordinary humanity, where we usually are, there is confusion and chaos. We have conflicting desires and our compulsions are in constant flux; there is nothing abiding, nothing that could be properly called 'self'. Only deeper down is there the possibility of increased understanding, integration and wholeness. Only in a state of collectedness, composure, openness and alertness can we know anything objectively. In all other states our perceiving apparatus is out of tune and it introduces its own noise arising out of the internal or external distractions and afflictions. Nothing that we decipher in these dispersed states is ultimately trustworthy. One of the purposes of the various spiritual schools is to help human beings repair themselves so that they may gradually come to a state of preparedness, freed of subjective desires, expectations, and fears. Only when quietened within may we encounter what is real.

In a purified and integrated state, arrived at after a long and arduous spiritual journey, we can discover our deepest self, called by various names by the sages: 'the spirit of the soul' (Eckhart), *acumen mentis* (Hugh of St. Victor), 'centre of the soul' (St. Teresa of Avila), 'spark of the soul' (St. Jerome), 'the divine person who is beyond the beyond' *(Mundaka Upanishad)*.

Healing of the soul is both a finding of the soul and its refinement and right ordering, an attunement to what is subtle and real and on which the manifest world is evidently dependent. Transformative self-knowledge is thus both a discovering and a sculpting of the soul. To say that human beings are made in the image of God is a statement of the real, not of the actual. For the actual to mirror the real needs work and spiritual discipline. In the process, the soul is healed as it finds its proper place in the Whole.

Healing of the soul quite naturally leads to the healing of our planet, in fact of the whole cosmos. There is an almost universal traditional idea of a correspondence between a human being and the cosmos, the microcosmos-macrocosmos homology. The fully developed person (*mahapurusha*) is said to mirror the whole cosmos. To become such a developed and whole person requires a transformation of the natural self. Then we discover not only that we are one with the cosmos in its physical and biological materiality and lawfulness, but also that the cosmos is one with us and has many inner and spiritual aspects.

In our attempts to find objective knowledge, the great aspiration of science, we cannot eliminate the person or the soul. What is needed is an enlargement of the person - freed from the merely personal and subjective - to be inclusive. This does not entail a horizontal extension of more and more knowledge, but depends upon a vertical transformation which allows a participation in the universal being.

The spiritual yoga - whether of the East or of the West (for Spirit by definition is non-sectarian and inclusive) - aims at the realisation of the First Person Universal. Such a person loves what they know. Only such a person can know without opposition and separation, freed from any desire to control or to manipulate. A person whose soul is healed, that is to say is whole, is naturally a sage.

Apart from the selflessness (and the accompanying absence of pride) and the natural feelings of compassion and love which are characteristic of all sages, there is one feature which needs to be underscored, and which is rarely remarked upon. A sage simultaneously sees the *oneness* of all there is and the *uniqueness* of everything. He is aware that each human being is a manifestation of One Divine Energy, but that at the same time each person presents a unique potential (and corresponding particular difficulties) and is a wondrously unique expression of the Vastness. Everyone is seen by the sage as both one with the Source as well as uniquely oneself.

Environmentally Conscious Citizens

Mauro Morelli

I am not a theologian or an economist, but a pastor who has learned not to worry so much about a place in heaven for the children born today but about whether they will have a place in our world and what kind of life they will have.

Over the last 22 years I have worked on the outskirts of São Paulo and Rio de Janeiro, the two largest Brazilian cities, an environment of pollution, violence and human destitution. Almost 1,000 people are killed every year in the two municipalities of the diocese and I am convinced that while misery and democracy are incompatible, degradation of environment and misery walk together as twin sisters.

Environment and social exclusion

Civilisation is the process of improving the quality of human life through the transformation of natural resources into food, housing, clothing, means of transportation, leisure facilities and provision of public services such as education, health, sanitation and water supply. Recently humankind has started to be concerned about sustainability or the impact and consequences of civilisation on nature.

We should have the same concern about the impact of progress on people. Throughout history, suffering imposed by progress has never been equally distributed, the greatest burden being placed on the shoulders of those who are socially, politically or economically weak. In the Middle Ages, for example, while the clergy and the nobles were enjoying the pleasures of life inside their castles and palaces, poor peasants lived amidst dirt and sewage, like the inhabitants of today's urban slums. In the last century, while business people made fortunes selling a great variety of products, the workers who manufactured them were dying of tuberculosis because of overwork, malnutrition, exposure to contamination and poor living conditions.

Today, while a minority of rich people enjoy a sophisticated and excellent quality of life, one out of six human beings goes to bed hungry every night and about 50% of the world's population are excluded from the benefits of economic growth and the development process. Many people suffer the consequences of accelerated urbanisation, such as lack of urban infrastructure, inadequate nutrition, educational and health services and exposure to chemical and nuclear pollution. In some places we are heading towards ecological disaster. It seems that we have come to such a degree of environmental and human degradation that people and even nations are treated as garbage that is not even worth recycling.

Recent trends from powerful economic groups point to the total liberalisation of international trade as the magic solution for poverty. Yet we watch with dismay and indignation an ever-increasing distance between the rich and the poor, with greater concentrations of wealth and misery and about 20% of all human beings consuming close to 75% of everything that is produced.

Even if the present economic system brought to every human being the average living conditions enjoyed by most people in the first world, the planet would be heading towards unsustainability. Mother Nature cannot sustain any longer the degree of consumption, spoiling and destruction that she has endured over the last decades. Governments and people need to work together to meet the needs and rights of all citizens and new mechanisms must be sought to guarantee citizens' control over state, public services and the market, which left to its own devices will never solve the problem of hunger and will instead accelerate the destruction of our beautiful planet.

If we truly want to build a world based on solidarity among people and nations and respectful to Mother Earth, we must look for new paradigms for progress and development. The victims of social and economic exclusion cannot wait for better days to come. This is our biggest challenge. We need to create the proper conditions to achieve growth and maturity to become partners in the history of our community. Any present or future goals must confront the sufferings of those who experience hunger and marginalisation. No argument can justify the denial of human freedom, peace and happiness to people living today. They cannot be mocked by the promise that their children will live in a prosperous future.

We have to advance inspired by great visions:

> Our vision is of people of different faiths beginning to learn from each other's spirituality and inspiration while developing practical examples in commitment to community and sharing. Our vision is of those with enough material goods beginning to live with less while replacing their idolatry of consumerism with a new spirituality, and of those with economic and political power making decisions based on the needs of all creation, leading to a fuller life for all. Our vision is one in which the local communities are empowered to resist the many threats to our survival.[1]

In this context, solidarity among people becomes the energy and source of social, economic and political transformation. It becomes a bridge that unites people while they share bread. Solidarity is a good way to restore human dignity to those who suffer starvation or live in affluence. The humiliation of any human being is a denial of everyone's dignity.

The path to a new civilisation

The crisis which disturbs the nations around the Black Sea and in many other regions of our planet has deep and ancient roots. We will need much wisdom, courage, humility and patience to find and follow the paths which will bring peace and guarantee life with justice and dignity for all people and nations. New technical solutions, new social habits and new cultural values can only be implanted slowly. Without a doubt, it is much easier to record a crisis, than to find a solution for it.

Crises are always accompanied by the seeds of renovation. The planetary dimension of the present crisis will contribute to the rise of new patterns of production and consumption. Even more, it will make us search for a new international economic order. How can any country remain prosperous and at peace while one billion human beings are condemned to languish in deprivation, suffering discrimination and the denial of their own dignity?

We are being forced to become aware of the limits of nature and of the rightful autonomy of nations. The new moral imperative is that the solutions for which we are searching cannot exclude any part of humanity. The need to eliminate hunger precedes politics: it is a basic human need and right. I believe that the self-confidence with which the First World countries impose and sustain their linear model of development is being shaken and the limits for development, whose existence was once denied, are now the subject of discussion.

There is an urgent need to implant a new model of development which does justice to the hungry and to the excluded of the human family and which respects the integrity of creation. The non-governmental organisations and religious institutions could assume the task of producing a minimal body of universal ethical norms, founded not in diplomatic bureaucracy, but on international networks in defence of fundamental principles. In this way we will be able to find new patterns of production and consumption. Domination will be replaced by respect and affection for creation. Bureaucracy and stagnation will be replaced by dynamism in cultural and social relations. Individuals deformed by greed and power will be replaced by people rich in solidarity and communion.

It is true that economy is not controlled by politics and neither are based on ethical values. The reality of hunger, for example, is neither talked nor thought about. The aim of the process of production does not include supplying basic necessities, only guaranteeing and increasing production, usually for export markets. For this reason, the National Conference of Bishops of Brazil, in a pronouncement made before UNCED, criticised the basis and the concepts of the proposition of sustainable development:

> Even though it is meant to have an all-embracing effect, reorienting the relations between peoples and between people and nature, sustainable

> development is still thought out within the economic sphere and it is within this sphere that the social element is thought of, the only new factor being the incorporation of nature in production costs. This reduction of nature to the purely economic is unacceptable. It is much more important to recognise the inherent value of creation than to put a material value on nature.[2]

For a model of development to be genuinely sustainable, visions which reduce social life to its economic dimensions ought not prevail. If a solution is to be found which will effectively preserve life on the planet, economic ideas will have to be submitted to ethical judgement. Ethics has as its object the analysis of the nature and purpose of our acts. Any definition of social ends which does not wish to be submitted to an ethical judgement is no more than an authoritarian imposition of hidden private goals.

To act ethically is to act with justice. Justice is, simultaneously, the central axis of all moral virtues in private life and the foundation of the ethical norms of public life. On the personal level, the virtue of justice is the internal and subjective disposition which leads a person to fulfil the actions prescribed by law and, through this, to become his or her own master. On the social level, justice furnishes the juridical and political norms which organise social life, guide the correct division of material goods and enforce respect for the common good. Therefore, to have justice as the basis of development, two principles ought to be recognised and respected:

- everybody possesses certain natural rights, rights resulting from the very essence of being human;
- the full and universal citizenship of all people must be assured through just laws and governments which provide and guarantee the basic necessities of the human person.

The first principle recognises the potential of citizenship; the second promotes the practice of citizenship. The application of these two principles does not mean the elimination of all differences due to injustice. Such differences are not metaphysical (first principle), but historical (second principle); therefore the challenge is not to eliminate inequalities, but to place all on the same level of justice.

Finally, it is my understanding that neo-liberalism does not meet ethical demands and suffers from the same evil which it condemns in Marxist economic proposals. Reflection brings me to the conclusion that the actual models of production and consumption of goods derived from nature are anti-ethical.

Faced with the devastation of nature and the increasing poverty and exclusion of millions of people from the Banquet of Life, I would dare to affirm that the model of development of the international economic order is not only unsustainable, but is also criminal and blasphemous. It tears the roots of life and eliminates hope from the face of the Earth.

Notes

1. World Council of Churches, (1991) 7th Assembly, Canberra, Sub-theme: 'Giver of Life - Sustain Your Creation', p.59.
2. *Igreja e a Questão Ecológica,* CNBB (1992).

The Moral Responsibilities of a Precautionary Approach

Zaki Badawi

Let me begin by committing a heresy. This, I hope, will allow me enough time to repent and return to the fold.

From the religious corner I come to praise science not to bury it. The benefits that scientists have given to the world are too obvious and numerous to need elaboration. They have equipped us with a power that has exceeded the imagination of the authors of the *Arabian Nights* and the Greek mythologies. Perhaps we have not always used it wisely. The blame can hardly be placed at the door of science. You cannot convict the manufacturer of the knife for its criminal use by a murderer!

There is a seemingly justifiable complaint regarding the methodology adopted by modern science in its tendency to fragment physical phenomena into manageable bits for ease of study which leads to a narrow outlook emphasising the part at the expense of the whole. The absence of a holistic approach has been forcefully criticised by eminent scientists at this symposium. I have often read and listened to vigorous onslaughts against modern scientific method and found the exercise long on diagnosis but short on cure. Could this be because the limits of scientific method are a reflection of the limits of the human mind and the span of perception? The phenomenal success of this method must surely justify its use.

There is some self-deception in the campaign against modern science. The criticism seems to assume the existence of an age of sounder knowledge and more successful procedure of inquiry in which the global outlook dominated and the fragmentation approach was shunned. In this idyllic age there was harmony between science and religion, peace and tranquillity prevailed, and the lambs felt secure in the embrace of the lions.

This mythical vision of the past needs to be challenged. It resides deep in our subconscious. The world before the scientific age was hardly more harmonious than it is now. Its most important characteristic was the dominance of religion. There is an illusion in some circles that religion is the harbinger of peace and tranquillity, of responsibility and care for others; if science has a criminal record, religions are not without blood on their hands. Remember the pious crusaders who sacked Constantinople. We do not even have to travel back to the eleventh and twelfth centuries; it suffices to glance at our daily papers or to listen to the news bulletins. From the Philippines to Sri Lanka, India, Pakistan, Afghanistan, Northern Ireland, Bosnia and

Palestine, religion is invoked to justify the shedding of blood, ethnic cleansing and the obliteration of the heritage of others in cultural massacres no different from genocide.

We of the religious fraternity have a lot to answer for and we must put our own house in order before we undertake to set the scientific world aright. It is the duty of the leaders of all traditions to alert their followers to the supreme importance of the need for all religions to cooperate with each other if they are to play a significant part in saving the environment. All religious systems are experiencing a period of revival which is felt not only in the major world religions but also in the traditional religions of Africa and among the original inhabitants of the Americas and Australasia.

This religious revival is probably due in part to the scientific revolution whose achievements have dwarfed mankind and estranged us from our environment. Scientific progress is the contemporary Frankenstein's monster. It revolves on its own axis driving humanity into investing ever greater resources to discover and invent. The terror of the atom bomb did not hamper development of the much more terrible hydrogen bomb and though the advances in genetics have prompted world rulers to set limits to the activities of scientists in this field, this may have been an exercise in futility, as the history of modern science is littered with discarded documents aimed at deterring researchers from penetrating into this area or that. The only means of directing the march of science is the cultural values which guide the choices of fields of study and help formulate the questions to be answered.

At this point I would like to paddle back into my own group. The failure of modern science was not due to its method. It was largely the result of its arrogant expulsion of faith and value from its enterprise. Teleology became taboo and the scientists paraded their wares as value-free. Perhaps there were justifiable reasons for scientists to seek to free themselves from the shackles of clerical fraternity but they were probably less wise than the emperors who rejected the dominance of the Church but retained their faith, Christianity.

Freed from religion and morality, science resembled a highly powerful speed-boat which moved at full power but without a rudder. It covered huge areas fast but without clear direction. Scientists then claimed for their discipline, and consequently for themselves, an unquestioned value. They became the measure of themselves just like the clerics whom they earlier despised.

The absence of religious faith and moral commitment has landed scientists in a maelstrom of problems. Often when they give answers they raise more questions and when they solve a problem they create more problems. It is time for our Sustainer to be invited to the scientific/industrial banquet; not as a guest, however honoured, but as at least a co-host whose vital contribution to steer the whole enterprise in the most sound and fruitful direction is sought and valued.

Islam, my faith, has from its inception placed a great emphasis on knowledge as the proper foundation for meaningful action. This principle is deeply embedded into our perception. The story of Genesis in the Qur'an does not identify the forbidden fruit as that of knowledge but of eternity. Satan in the Qur'an whispered to Adam and his spouse, 'Shall I show you the tree of eternity and never ending dominion?' Both were thus enticed with the promise of eternal life. Not for us the association between innocence and ignorance. The Qur'an does not portray Adam as a henpecked husband nor his wife as a foolish temptress; nor does it accord the dumb snake such persuasive cunning.

Had the myth of Prometheus been Islamised this kind bringer of light and knowledge would have been honoured not tortured. The distinguished American historian, Marshall Hodgson, observed that had modern science developed in a Muslim environment, it would have borne a collective and spiritual mark in contrast to the western individualistic and materialistic kind.

Now let us come to Islam's view of our place in the world and our relationship with it. Whether we gradually arose from the amoeba in a long process of evolution or whether the Creator telescoped the evolutionary stages in one single step makes no difference. From the late nineteenth century both views have found adherents among Muslim theologians. They assumed total concordance between faith and knowledge. To them the Qur'an is the word of God, and the world is the work of God. It was to them unthinkable that the work of God would contradict His word. Their pious eyes could observe the fingerprint of Allah in the fragrant flower and the dead leaf, the flying birds and the wild beast. Even human inventions were placed among the miracles of the Creator Who in His Wisdom and Mercy has endowed this particular species with such skills. They looked upon the world with awe and veneration and sought to enjoy but not to master its creatures.

Human beings' relationship to this world resembled in many ways a form of property contract which I learned about for the first time in England, where there are two types of ownership: freehold and leasehold. The freeholder has no time limit and has absolute power over the property. He can do with it what he wills. On the other hand, leasehold is limited in time and accords the user the right to use the building with the responsibility for its proper maintenance only until handing it over in good order at the end of the lease.

Our dominion over the world is of the leasehold type. We are therefore not entitled to tamper with it without the permission of the owner of the freehold, God. How do we obtain such authority? Islam tells us that authority is enshrined in the Shariah, the Divine Law. There, the rights and duties of us the lessees are clearly marked. We are to care for our habitat and protect it lest it decays. The Shariah never loses sight of the need to guide us along the path of care for the environment. This is demonstrated most clearly in the Islamic rules of engagement. War is a most hated act to a faith as committed to peace as Islam; but Islam accepts that conflict can and does take place. Rules therefore have to be set for it. The Islamic rules of war are testament of a great concern for the environment. The Shariah orders warriors not to desecrate a place of worship nor to harm a peaceable man of religion, nor the old nor the wounded nor the infirm. The warrior must protect women and children.

The rules also demand respect for the environment. Fruit trees must not be damaged, water resources must not be polluted, every asset of social or commercial value must be fully preserved. No mindless damage to the environment is allowed. Improvements are permitted as long as they adhere to the prescribed guidance. The Shariah prohibits waste and extravagance and consumption has to be moderate. The wastrel and extravagant can be deprived of freedom of action and placed under the control of a guardian.

According to the Shariah, no benefit should be obtained at the price of inflicting harm on the environment. Safeguarding against polluting a water source, for example, would take precedence in Muslim agricultural planning over growth in food production. On the other hand, cutting down fruit trees to stop the spread of wild fire is permissible.

Islam does not concern itself only with prevention. It instructs us to replenish the environment and improve it. This responsibility has to be discharged even under extreme conditions. Prince Sadruddin Aga Khan quoted a saying of the Prophet: 'On doomsday if you have a sapling and you have just time enough to plant it then do so'.[1] To explain the full significance of this saying let me refer you to our eschatology. We believe that at the end of time the angel Israphael will blow a horn which will be heard throughout the world to warn all living beings that the end is nigh. Soon after, the second blowing of the horn comes and every living thing dies. In between the two blasts of horn, when everyone is certain of death, the Prophet tells his followers that under these circumstances they should discharge their duty to the environment by planting the saplings held in their hands. Such is Islam's commitment to the environment.

Note

1. See Chapter 2.

To Till It and To Keep It: on Ecology and Everyday Life

Naftali Rothenberg

Interreligious relations are one of the most problematic and complex areas in the history of civilisation. The major denominations thrived as a result of the lack of cooperation between them and interreligious tension has in retrospect made a fairly sizable contribution to the creativity and cultural uniqueness of the various religions and sects. Yet there may exist just one pursuit on which diverse religious leaders and believers can agree: preventing the destruction of the world in which we live. This is not only a search for consensus, but also for practical influence on human behaviour. Despite the many diverging opinions and doctrinal disputes, the influence of religion can ignite change in the individual's attitude to the environment and, in effect, in his attitude to himself.

The intense and sometimes obsessive concern with environmental protection is considered a post-modern phenomenon that tries (and often pretends) to repair a world destroyed by processes resulting from the expansion of modern life. Many are astonished to discover that the religions, in their source material, are overflowing with ecological doctrines constructed and developed in the areas of ideology, theology and philosophy, as well as in the practical aspects of environmental protection.

Religious viewpoints that offer a balanced and harmonious approach can have great influence. The foundation of these viewpoints is in the Bible: 'And the Lord God took the man, and put him into the garden of Eden to till it and to keep it' (Genesis 2:15). The connection between these two words, 'to till' and 'to keep,' expresses the essence of balance, the middle road that facilitates harmony between man and his world, a path to a solution that is both ethical and practical. Numerous commentators from various religions assert that this divine command is not limited to the Garden of Eden. God transferred his world to human control and for human enjoyment, but only on condition that we maintain it properly and adopt an ecologically sound lifestyle.

In Ecclesiastes we read: 'I know that, whatever God does, it shall be forever: nothing can be added to it, nor anything taken from it...'(3:14) and: 'Consider the work of God: for who can make that straight, which he has made crooked?' (7:13).

There is a discussion of these verses in an ancient Jewish commentary, the Midrash:

> When the Holy One, Blessed Be He, created Adam, he took him and led him past all the trees in the Garden of Eden, saying: Look how pleasant

> and praiseworthy are my deeds; everything I created, I did so for you. Be conscious of this, so that you do not ruin and destroy my world, because if you ruin it, there will be no one after you to fix it (Kohelet Rabbah).

Today's alienation from religion and its messages and from life in harmony with nature goes hand-in-hand with the development of a modern idolatry that believes in the unlimited power of the human race and enables humanity to ignore not only issues between humans and God, but also ethical questions relating to the interface between people and their environment.

In post-modern ecological world-views, there are two extreme approaches at either end of the spectrum. On one side is the outlook that sees the human being as lord over nature and justifies the subjugation of nature and any use of it to advance the needs, survival, and enjoyment of human society; on the other side are those who see nature and its protection as its own purpose and humans as merely one of the elements within it, equal in value to the rest of the elements but different in the degree of destructiveness. These positions reduce the chances of progress in environmental protection and perhaps also cause alienation among wide sectors of the population.

The religions present a third way: the world was created for humans, who are at the centre of the scheme, but they do not have license to be aggressive conquerors because the world does not really belong to them but rather to God. The religious principle therefore is expressed in the placing of responsibility on humankind to find the balance between 'tilling it' - intervening in nature in order to survive - and 'keeping it'- avoiding causing destruction and ruin that cannot be undone. I believe that, on a practical level, this moderate and balanced approach has a chance of becoming accepted by the very broad segments of the world community that do not accept the radicalism of some of the Green movements but yet are concerned by damage to the environment.

The ideology, philosophy and ecological theology of religions in general, and Christianity in particular, are based on belief in a Creator, on the idea that life has a purpose, and on the view that man has a task in this world and that human reality is not merely a statistic. The philosophical-theological approach is dominant today in the new popular literature on religion and environmental protection. The practical and behavioural approach to environmental protection, the sources of which are developed in the Shariah (the Islamic legal system based on the Koran) and in Halacha (the system of behavioural rules and laws in Judaism based on the Torah and Talmudic interpretation) are less well-known. Here we are dealing with a detailed system of laws and regulations that addresses all aspects of environmental protection and sets forth practical rules of behaviour.

The practical approaches in Judaism and in Islam are the product of daily life and the real needs of people who take upon themselves ethical norms that produce, in effect, a balance with surrounding nature. These rules are difficult to distinguish from the environmental protection legislation in today's secular parliaments. Of course, it must be noted that contemporary legislation was formulated to provide a defence and a response to the violation of the balance in nature and to the destruction of the environment, while the rules of Halacha and Shariah are thousands of years old and were originally formulated so that we would not arrive at the sad situation in which we find ourselves today.

One of the most far-reaching protective measures that determines our relationship to the environment is the prohibition of senseless destruction.

> When thou shalt besiege a city a long time, in making war against it to take it, thou shalt not destroy its trees by forcing an axe against them: for thou mayst eat of them, and thou shalt not cut them down; for is the tree of the field a man, that it should be besieged by thee? (Deuteronomy 20:19).

Even in the extreme situation of war, that by its very nature runs roughshod over normal moral behaviour, the Torah demands consideration for the environment. Of course, Halacha and Shariah also prohibit the destruction of trees in times of peace and tranquillity. These two systems expanded this prohibition to include destruction of all vegetation, animals, and property, while assuring the integrity of the public domain.

The sages of the Halacha did not consider the public domain to be a no-man's land in which everyone can do as they please, as the following Talmudic anecdote makes clear:

> There was a man who would clear away rocks and then throw them from his private domain into the public domain. A righteous man came upon him and said to him, you senseless! Why are you clearing away rocks from a domain that is not yours to your own domain? The other man scorned him. After some time the man needed to sell his field. He walked through the very same public domain and stumbled over the very same rocks. He said: That righteous man was correct, why are you clearing away rocks from a domain that is not yours to your own domain? (Baba Kama 50b).

In other words, the private domain is sometimes ours and sometimes passes to others, but it is the public domain that always remains ours.

Another key area in the Halacha's ecological blueprint is the assurance of water quality. Clear regulations are laid down regarding society's and the individual's rights and responsibilities in this area. These regulations concern formation and development of water sources, the rights to their usage, the proper maintenance of reservoirs and other facilities, prevention of water pollution, and awareness of the dangers arising from it.

Other areas of concern are noise pollution in urban areas; city planning; industrial damage in urban and village environments; garbage disposal; balance between agriculture and natural vegetation; limits on fruit trees as compared with ornamental trees and the separation between them; treatment of animals; the prohibition of cruelty to animals, and more.

One realm in which we find a rich Halachic literature, evolving over more than 800 years, is air pollution and smoke damage. The very same problems that occupy humankind in the post-modern era were already being discussed by the sages 1,000 to 1,500 years ago. The health and welfare of the individual take precedence over other factors; there must be due consideration of the individual's quality of life; there must be elimination and prevention not only of things that cause damage to health, but also sources of irritation and distress. The Halacha deals with the tension between, on one hand, the damage caused to the individual or group from foul odours or smoke, and, on the other, the dilemma inherent in wiping out a source of income. Not in every case does the Halacha recommend erasing the source of pollution; it often protects the livelihood of local residents, even at the cost of ecological harm. This is the question of balance alluded to earlier.

We should be publicising this surprising match between progressive environmental protection legislation and the regulations of the Shariah and Halacha. Many Muslims and Jews are not aware of the classic ecological regulations contained in their own religions. Similarly, many Christians never experience a theological discussion of the harmony between man and nature. Traditional and devout individuals from all religions attribute awareness of environmental protection to the secular culture and do not see it as part of their tradition or their religious obligations. They have a sceptical attitude toward some of their co-religionists who are involved with ecology and suspect them of divorcing themselves from the 'pure' heart of religion. Anyone who is familiar with Shariah and Halachic sources and with the Book of Revelation and all the sacred texts from Genesis forward knows that there is no reason to be apprehensive. Today, it is becoming increasingly clear that commitment to environmental quality is anchored deep within the sources of all the religions. Each religion can mobilise its forces on behalf of this commitment, highlight it according to its own principles of belief and action, and influence the internal dialogue in other religions and cultures.

Ecological commitment is not synonymous with modern-day 'nature religions.' Ecological doctrine in the major religions relates not only to the interface between humans and the environment, but also to the relationship between people and between individuals and themselves. Ecological ethics is indivisible: those who direct all the intensity of their compassion towards animals; who labour to maintain beauty and aesthetics only for the natural environment; who think that it is possible to deal with polluters of the Black Sea without critically examining all aspects of human behaviour; all these lack a complete moral system.

As we deal with these topics, learning, teaching, publicising and responding; as we support environmental protection legislation and enforcement; as the general public accepts an ecological outlook as the ethical and religious norm common to many religious cultures, and does not identify such a viewpoint with secular radicalism; as we link and synthesise environmental values into traditional cultures we shall then magnify the real, practical opportunity to live in a higher quality environment and in a more balanced and well-maintained natural setting. This is the way to realise the principles of morality and faith while blending proper, balanced, and ecologically-minded behaviour into our everyday lives.

Living in balance with nature and the environment demands from every person a continuous effort. It is not a vain endeavour, but rather a path to human development and a way to draw closer to God. It raises the status of man to the heights of 'a little lower than the Lord':

> To the chief musician on the Gittit, a psalm of David. O Lord our Lord, how majestic is thy name in all the earth; who hast set thy glory above the heavens. Out of the mouth of babes and sucklings hast thou founded strength because of thy enemies, that thou mightest still the enemy and the avenger. When I behold thy heavens, the work of thy fingers, the moon and the stars, which thou hast ordained; what is man, that thou art mindful of him? And the son of man, that thou visitest him? Yet thou hast made him a little lower than the Lord, and thou dost crown him with glory and honour. Thou makest him to have dominion over the works of thy hands; thou has put all things under his feet: all sheep, and oxen, and also the beasts of the field; the birds of the sky, and the fish of the sea; whatever passes through the paths of the seas. O Lord our Lord, how majestic is thy name in all the earth! (Psalm 8).

Human Health and the Environment in the Black Sea Region

William K. Reilly

The health impacts of the vast and continuous insult to the environment of the Black Sea countries are considerable and water contamination is but one, though I think the most important, of the problems facing the region.

Water in the Black Sea basin is characterised by high levels of fecal coliform, copper, nutrients such as phosphates and nitrates, phenols, hydrocarbon and petroleum waste. Although the Black Sea is less saline than the Mediterranean, damming upstream has resulted in reduced water flows and increased salinity. Heavy sewage loads, of human and industrial waste, and pesticide levels that measure as much as 5,000% to 10,000% greater than World Health Organisation guidelines, are found in coastal areas of Russia.

Toxic, viral and bacteriological diseases occur with a frequency vastly exceeding western European averages. These incidents of acute illness are exacerbated by regional shortages of medicine and medical care. There is a serious risk of disease contagion and mutation outside the region. Reported data for the Black Sea region (and there is reason to believe that under-reporting is common) indicate that common diseases include diptheria, typhoid fever, cholera, hepatitis A, hepatitis B, polio, all childhood diseases, dysentery, and parasites of the eyes, ears and digestive system. Radiation-related cancers also exceed western European averages. Not all, but most of these diseases are water-borne.

The groundwater in the region is particularly affected by high rates of pesticide use. Excessive rates of pesticide application have been common in the region, though recently their use here has declined because of the economic downturn. In Russia, heavy pesticide use has been associated with an increased frequency of diabetes, anaemia, kidney and thyroid conditions, and rising levels of infant mortality and morbidity. In the Black Sea region of Russia, the levels of these illnesses exceed by three to ten times those prevailing in Russia as a whole.

The day-to-day reality of people affected by contaminated water is often not obvious or dramatic, but it can be grim. I recall visiting an Eastern European city while I was Environmental Protection Agency Administrator and being warned under no circumstances even to brush my teeth in the water. I asked a young member of our delegation who had recently spent a year in that city how the locals coped with their bad water. Did they avoid it, I asked, or boil it, or develop an immunity? He told me that everyone he knew well enough to know about such things had chronic diarrhoea.

One person has told me of vacationing in the summer at Sochi, a famous Black Sea health resort visited by Boris Yeltsin. Every summer, she says, she got dysentery. When she tried to learn how common the affliction was, she was told that there was no record kept of hospital admissions for dysentery. She noted, however, that the hospital was full to capacity with dysentery patients.

There is no shortage of water in the Black Sea region. Water consumption here is actually higher by far than in many areas of western Europe - some 2,000 cubic meters per capita, or about the same level as in the United States. A water-rich area has a greater capacity to address water contamination problems, but perhaps the need to clean and recycle water, thereby preventing polluting discharges into the sea, is less obvious and immediate than where water is in short supply.

Surveys in the region indicate that large amounts of wastewater are totally untreated. Efficiency of wastewater treatment plants is under 70%. Less than half of the wastewater that *is* treated is treated to secondary treatment levels, that is, to a point where 85% to 90% of contaminants are removed. Wastewater treatment plants are poorly designed, have outdated equipment and inadequate systems of monitoring and control. Metallurgy and chemical industries are a major component of Ukraine's economy. Both sectors generate large amounts of hazardous waste. Less than 30% of this waste is recycled. Only 1% is rendered safe. Some 40% is deposited in uncontrolled landfills. Is it just a coincidence that drinking water in Ukraine has significantly deteriorated in recent years?

Twenty-five years ago in the United States, two thirds of our surface waters were unfit for swimming or fishing. One river, in an industrial area of the state of Ohio, even caught fire. Ten years after the Congress, prompted by the Ohio and other incidents, committed the United States to 'fishable and swimmable' waters everywhere, my predecessor as EPA Administrator took stock of the situation and commented that, while we could not yet claim that all of our rivers and lakes were fishable and swimmable, at least we could take pride in the fact that they were no longer flammable.

As a former government official, I am moved to ask, what would I do if I were responsible for the environment here? Where would I put my priorities, and how would I engage public and official support?

I would begin by recognising the need for significant investments in water purification and wastewater treatment. Some of the capital to finance this infrastructure will necessarily come from the region itself. Some of the needed financial resources may come as aid from national governments and multilateral aid and lending institutions like the World Bank and European Bank for Reconstruction and Development, which often finance needs surveys and planning studies. Drawing on the experiences of cities such as Buenos Aires and Manila, it is even possible to have expensive and state-of-the-art systems for collection, treatment, purification and distribution of water entirely financed by private investors. In the latter case a government agency, acting with the advice of a bank or international agency such as the International Finance Corporation, conducts an auction to grant a sale or long-term concession to a consortium of engineering and construction companies and water providers. Bids are entertained from pre-qualified groups and the commitment required is to detailed performance improvement and a specific tariff that the bidder proposes to charge the consumer.

In the case of both Buenos Aires and Manila, the tariffs agreed to by the winning bidders are actually significantly lower than the water tariffs in effect at the time of the auction. The bidders simply factored in the savings obtainable from improving bill collection, reducing waste, and downsizing bloated public agency staffs. Those reductions were large enough to finance major investments in new technology and equipment, and still permit a substantial drop in tariffs to consumers - 22% in Buenos Aires and more than 50% in Manila. Leakage in the public water supply of Sofia, Bulgaria, is 60%; there is clearly potential for improvement here.

Certain preconditions are necessary to attract private investors to finance public infrastructure. There must, for example, be a commitment by government to enforce collection of tariffs. Where, as in Russia and Ukraine, there is a tradition of free water service, it is unlikely that the government would be willing or even perhaps capable of establishing a policy requiring water to be paid for. If so, then loans from the World Bank or EBRD are also unlikely to be forthcoming because loans, after all, do have to be paid back, and without water tariffs there will be no means to do so. Unfortunately, I see little prospect for upgrading infrastructure in such circumstances in the near term.

Conditions in Turkey and Romania appear more conducive to attracting investment in water infrastructure. New governments appear committed to expanding the role of markets and attracting private capital. Both countries have under-exploited tourist resources that could capture a far greater clientele if coastal waters were reliably clean and drinking water safe. Were these countries to demonstrate the efficacy of a strategy of systemic clean-up financed by private capital, the example could be compelling to other governments in the region.

The basic principle underlying private, or really and truly efficient and sustainable investment, is that tariffs must be sufficient to cover the costs of the service. Long years of under-capitalisation give rise to the poor maintenance and leaks evident in systems today. Competition for public subsidies left the infrastructure for water short of basic operating needs, and woefully deprived of capital for improvements and upgrades. Some eastern European countries have managed their transition to market-based economies in a way that benefited the environment. Both Poland and the Czech Republic have improved their environment. One of the ways this was achieved was to allow water tariffs to rise 150%, the steepest increase experienced in the region. Those tariffs are permitting new investments to be made and debt to be incurred and serviced.

Financial aid to emerging markets and developing countries from aid agencies of governments and multilateral institutions has been declining throughout the 1990s. Last year the total of such aid was $44 billion, roughly the same as in 1990. Yet, as the Czech, Polish, Argentinian and Philippine experience demonstrates, it is possible to attract private funds to support environmental clean-up. Flows from developed to developing countries of private investment, which were roughly equal to public governmental transfers in 1990, last year reached nearly $250 billion, more than 15 times the amount of public aid. This is the flow of capital that offers promise, that is growing, and that can and must be attracted to the environmental cause.

Governments in this region cannot, however, sit back and wait for someone else to carry out the clean-up. Government vigour is essential to set priorities, prosecute environmental crimes and to ensure that corrosive pollutants from chemical or metallurgical facilities are pre-treated before entering the public system. Governments must monitor and inspect, set and enforce performance standards, ensure transparency in contracting or overseeing bidding, and communicate to farmers and others the practices important to reducing pollutant stresses on groundwater and keeping runoff and discharges to water bodies to a minimum. When a government downgrades its environment ministry, as Russia did in 1996, it sends a signal that it is not serious about the environmental priority. Investors, public and private, will not ignore that. Moreover, when governments lower the proportion of gross domestic product allocated to the environment, as Bulgaria has done, from 0.93% in 1989 to 0.29% in 1993, that too will be noticed and interpreted as an indication that the government has not yet recognised the urgency of a catastrophic environmental condition. Until it does, outside investors would be well advised to plough more fertile ground.

I once directed a programme aimed at reducing the contaminants reaching America's largest estuary, Chesapeake Bay, and in which one of the most intractable problems was to persuade farmers in Pennsylvania, far removed upriver from the Bay, that the fertilisers and cattle waste their farms generated were responsible for algae blooms and oxygen depletion in the Bay. The Black Sea drains a third of Europe, home to 160 million people, many of whom discharge their wastes far upriver from the Sea, and the challenge of informing and motivating a public, collective response is a formidable one.

In my experience, arguments based on purely rational and scientific calculations rarely stir people to action, but spiritual and religious appeals can and do engage the heart and fire the soul. Religion informs morality and maintaining the natural systems that sustain life is fundamentally a moral issue. Bringing the religious perspective to bear on the Black Sea is an effort to enlist the emotional sympathies and spiritual attachments vital to mobilising people's concern and action. It is, I suppose, about love. Jacques Cousteau, that eminent pioneer of the sea, once said 'In the end, people protect only that which they truly love'.

Environmental Equity

Severn Cullis-Suzuki

I do not have a degree in science. I cannot tell you much about how our economy works. I do not even have a job. I am 17 years old and have just graduated from high school. What I can offer, however, is the perspective of youth.

When I was 11, over six years ago, I heard about the Earth Summit to be held in Rio in 1992 and decided I was going to attend. It was to be the largest meeting of political leaders and heads of state ever assembled and I wanted to go to remind them of the reason they were there: their children. I had started a small children's group with some of my friends called the Environmental Children's Organisation (ECO) and for a year we fundraised and campaigned and, thanks to luck and support from a lot of people who believed in our idea we made it.

Too often children and young people are not included in conferences. I met only two children who were official delegates at Rio. Without them it is easy to forget that it is children who have the biggest stake in those sessions. Because I was so young, I did not understand how complicated the world of politics and economics is. I saw the world in a very simple, straightforward way. Now, after five years of attending conferences and giving many speeches I have become more aware of the complexities that decision-makers have to deal with, but the simplicity of a 12 year old's perspective remains undeniable. Even though I am now 17 (and over the hill), I still know that the bottom line is - my future is being compromised.

To me, environmental equity means generational equity. In the past, every generation worked to improve the prospects for their children, to increase their opportunities to lead a richer life. Today it is apparent that my generation will not get the same, clean legacy that was handed down to our parents.

Young people in North America and, I believe, all over the world, are concerned. We have reason to be. Our future is in danger and with today's media focus on global crises, it is difficult for children to remain oblivious to what is happening to our world.

Human beings have been altering our environment since the beginning. We have been logging and mining and fishing for thousands of years. So why is it suddenly such a problem? It is because all of a sudden the scale of alteration has been massively amplified, and the results have been huge. According to a report from the Worldwatch Institute, most of the destruction of forests has taken place after 1950. The vast bulk of ecological change that has occurred over the last 10,000 years has happened within my parents' lifetime. If you talk to any fisherman or forester over the age of 40 you will hear countless tales of how 'it used to be'.

I know you must have seen changes in your own lifetime, because even I have! When I was young, there was an annual fishing derby in Vancouver where people would catch giant chinook salmon. About ten years ago, the derby was cancelled because the chinook had become extinct. The view of the mountainside from my grandparents' cabin where I spent my summers is now eaten up by gravel pits, subdivisions, and clearcuts.

I have just come back from spending a month in one of the most beautiful reserves of nature I have ever seen or imagined. I lived with a Haida family who took me out and shared with me the bounty of their home, Haida Gwaii, also known as the Queen Charlotte Islands, a little strand of islands off the west coast of Canada.

Every morning we would go out food gathering to the reefs at low tide and spear goodingaay, the sea urchins. Before we collected some in our buckets for relatives and friends we would sit down and crack the urchins open to feast right there on the rocks as the Haida have done for millennia. I was amazed and overjoyed to see the alguthian - the abalone which have been wiped out by commercial divers up and down our coast. We caught fish: we trolled for spring salmon, set skates for halibut, and jigged massive cod. I used to think of fishing as something that requires a lot of patience, but in Haida Gwaii, if you do not get a bite after five minutes you move or check your hook because something is wrong.

The foundation and basis of the communities of both Haida Gwaii and the Black Sea revolve around the food the waters provide. The plentiful halibut, cod and herring of the Haida remind us of when the Black Sea was thick with bonito, salmon and anchovy. The waters of Haida Gwaii are now beginning to be stripped as the Black Sea has been stripped. You do not have to be a scientist to recognise the possible fate of Haida Gwaii.

The Black Sea represents all of our Haida Gwaiis. We must learn from this sea and use it as a lesson, face up to what we are doing and look ahead to what we must do in our own backyards.

My mother once told me that children are closer to creation. We have not let go of the connection and love of nature with all of its puddles, tadpoles, flowers and furry creatures. We understand when native people say, these are our brothers and sisters. We are part of nature still.

The secret to rediscovering our real values is to remember what and who we were as children. Remember all the insects and birds, catching butterflies and looking for frogs in ponds? Remember playing in the grass and climbing trees? Remember how important they were, how you could not imagine your world without them?

What concerns me is that today's adults may be the last generation to have childhood memories of nature as it always has been.

Everyone can talk about the environment - we have the experts and the inspiration right here. It is now time to stop the talking and start the doing. I will turn 21 in the year 2000. I will spend all of my adulthood in the twenty-first century. My whole life is ahead of me and I will live with the consequences of decisions that are made today. I will inherit what you leave me.

Chapter 6

The Beauty of Coastal Landscapes

Introduction

Erdal Özhan

Coastal landscapes are the mirrors which reflect the health of the rich coastal ecosystems and the way human beings deal with them. Many natural features and a great variety of habitats are presented by coastal landscapes. Man-made additions to coastal landscapes indicate the cultural vigour of the maritime nations which have historically lived, or are presently living along the coast and the beauty of coastal landscapes has been an effective factor in the boom of coastal recreation and tourism. This chapter deals with the value of the Black Sea coastal landscapes, the perception of order and beauty in nature in general and of the landscape in particular, and tourism, an ever-growing industry which is much interrelated with coastal landscapes.

The connection of human beings to land and landscape is addressed by Neal Ascherson who points out that 'the relationship is not of master and slave but nearer to the traditional imagery which used to describe the relationship between men and women', and quotes a story of a traveller who asks the striking question 'How can a mountain *belong* to anyone?' Ascherson also draws attention to much religious tradition which presents the natural world, biotic or abiotic, as 'an estate to be managed by Man', and to be exploited 'either carefully or reverently, or recklessly and selfishly'. Ascherson poses the question: 'Has religious thought moved forward to a point at which it fully acknowledges the integrity of creation in all its forms?' He rightly argues that the old perspective of 'an enlightened human stewardship' of nature should be modified, but the new approach should not 'demolish the imperative to care for life-forms and landscapes'. While acknowledging the beauty of Crimea and its landscapes, he emphasises the importance of the Black Sea shores to many nations both historically and today, and urges that the Black Sea and its coast 'deserve reverence'.

Bishop Basil of Sergievo points out a parallelism between the thinking of Dionysius, the sixth century theologian, and the work of David Bohm, a contemporary quantum physicist, in the sense that both conceive of a 'layered' world. The universe is made up of an 'order of ever increasing orders of subtlety in the implicate order'. He refers to two important attributes: comprehensiveness and 'non-locality'. These are expressed by the statements that: 'In any understanding of any level, *everything* must be taken into account,' and 'What happens to one affects what happens to the other even though they appear to be unconnected'.

Bishop Basil points to the need to find 'a contemporary language with which to speak about the world as we know it through science and then relate this to our spiritual concerns and to God'. He further proposes that the language about the material world would be 'something like that of David Bohm, a language that opens out easily into the world of mind and spirit'. The beauty of the Black Sea coastal landscape 'is an objective property of nature' which is 'accessible and perceptible by every human being' and the 'ultimate source of this beauty is God - the source of all that is'. Bishop Basil concludes that 'our ecological task is to find ourselves in the universe - and to find the intellectual universe in us', and that 'our understanding will never reach the depths that are in us'.

Nicoloz Beroutchachvili outlines the advances that have been made in researching the landscape of the Black Sea countries and highlights the remarkable physical and ethno-linguistic diversity of the region. He too emphasises the connection between the landscape and those who inhabit it, describing how, 'just as man or an ethnic group can affect the shape of the landscape so ...the landscape itself affects the human being'. He explains how the history of the development of the Black Sea region reflects cultural perceptions that have altered over the centuries and how human activities, such as tourism and agriculture, have impacted on the landscape.

Terry De Lacy provides statistics about world tourism, the world's largest industry. He points out that environmental degradation results from the inability of the present market economy to value the environment properly, and that tourism has great potential for doing this. Tourism is described as being a relatively less consumptive form of land use compared with other industries. He provides guidelines for achieving 'ecologically sustainable' tourism, describes the policy instruments that can be used, and provides an example of best management practice for tourism development. He notes that according to a survey, about half of travellers appear to pay attention to the environmental awareness of travel companies, and that the same percentage would agree to pay more for environmental protection. He concludes that 'the environment has become mainstream - it is being internalised in all our professions, disciplines and industries'.

Land, Sea and Belonging

Neal Ascherson

Many people, individuals and families and nations, have been seized by the beauty of the peninsula of Crimea, and its many landscapes. Behind the rocky shore and ancient port-colonies is the coastal range, crowned with mountain fortresses, and behind that in turn an expanse of steppe, grey, windy and open; then to the east there is yet another countryside of steep downs leading out to the Kerch Straits - the Cimmerian Bosphorus, one of the many places held to divide Europe from Asia.

A journalist I know went to talk to a Crimean Tatar woman, a few miles away from Yalta towards Simferopol. She had returned from a generation of exile in Kazakhstan, inflicted on her entire people by Stalin in 1944; on a patch of gravelly waste where the local Russians and Ukrainians tolerated her presence, she was putting up a shack of boards and corrugated iron and chopping a vegetable patch into the earth. Asked why she was doing this, she summoned up a word for what was rarest and most sweet and satisfying in her world, and she replied: 'This land is my Snickers!' It did not matter that this was not the same patch that her ancestors had cultivated in the time of the Tatar Khans. The point was that she was back in Crimea, somewhere within sight of the peak of Shatir Dagh.

Men and women still risk and lose their lives to defend - as they put it - a certain stony hill whose outline is a background to their dreams, or a few hectares of what seems to others a stretch of featureless marsh. This love and identification that they feel is certainly about a kind of possessing but it has seldom led to a crude claim of total possession - not at least until the last century or so. Instead there has been a sense of holding a piece of land in trust, doing one's best for it to honour the privilege of occupation. To look up at a hill after returning from a journey, or to recognise the special way that a path turns among small birch trees - that is not a statement that 'This is *mine*'. It says something else: 'This is where I belong'. In a human being's connection to land and landscape, the relationship is not of master and slave but nearer to the traditional imagery which used to describe the relationship between men and women.

I make this point about Crimea, landscape and possession in my book, *Black Sea* (Vintage, 1996):

> All human beings are in some sense immigrants. All hostility between different cultures in one place has an aspect of the classic immigrant grudge against the next boatload approaching the shore. To defend one's homes and fields and ancestral graves against invasion seems a right. But to claim unique possession - to compound the fact of settlement with the

> appearance of a landscape into a single abstract of eternal and immutable ownership - is a joke.
>
> Crimea, whose beauty provokes almost sexual yearnings of possession in all its visitors, has demonstrated this joke in almost every century of its history. It has no natives, no aboriginals. Before the Scythians, before the Cimmerians who supposedly preceded them or the Bronze Age populations who raised the first burial mounds, there were human beings who had come from somewhere else. Crimea has always been a destination, the cliffs at the end of the sea or the shore where the wagons must end their journey across the oceans of the steppe. Voyaging communities settled in Crimea (the Scythians lived here for nearly 1,000 years), but in the end they all dispersed or moved on. All that has been constant in Crimean history has been a certain structure which the peninsula has imposed upon its visitors: the zones of mind, body and spirit have often been effaced but until now have always reemerged. Only in recent times has the Crimean truth - that it belongs to everybody and to nobody - been violated. Two of these violations, which would be merely absurd if they did not imply so much blood and suffering in the past and very probably in the future, are the declarations of two autocrats. In 1783, Catherine II proclaimed that the Crimean peninsula was henceforth and for all time to become Russian. And in 1954 Nikita Khrushchev, a Ukrainian seeking to divert the attention of his own people from their miseries, announced that he was transferring Crimea from Russia to become for all time Ukrainian.

The Scottish banker Iain Noble once described to me his own Saul-Paul experience on this matter of possession. Driving across the West Highlands, he had given a lift to a young Israeli hitchhiker and tried to entertain his passenger by identifying the landscape as they passed across it. 'That's Ben More - it belongs to the Duke of Atholl. And that's Loch Breac, it belongs to a Canadian millionaire. And down there you can see Glen Etive - all those hills belong to Fleming the banker'. He noticed that his passenger was staring at him. The Israeli asked; 'Excuse me, but I didn't understand. That's a mountain - how can a mountain *belong* to anyone?' Iain thought about this, and found that he had no answer. Soon afterwards he retired from banking, sold much of his property and opened a college of Gaelic culture on the Isle of Skye.

This beautiful landscape of Crimea which has inspired so much poetry and prose - Pushkin, Mickiewicz, Chekhov, Mandelstam, Tsvetayeva, to name a few names - and so much love and longing, which has been soaked in innocent blood so many times and lost so many exiles, is trying to tell us through its history a message which is central to this symposium. Before we can work though to a practical programme into which religious and scientific and even utilitarian perceptions can flow easily and make their proper contribution we must address the question of how human beings relate to the natural world. Are we apart from it, as the masters of a creation given to us on trust to exploit either carefully and reverently, or recklessly and selfishly? Or are we integral components of it, members of the natural world no more and no less than the migrating cranes and the anchovy shoals and the monk seals?

Much religious tradition, especially the Judaeo-Christian one, presents the natural world as an estate to be managed by Man, its creatures and flora the innocent serfs whose very being have a subordinate character. There is a passage in Genesis in which the Lord presents the animals to Adam and invites him to name each one (Genesis 2:19, 20). This passage is amazing at many levels, not least as a very modern parable about language in which existence follows rather than precedes the Word. It also shows naming as the critical act of taking possession - rather like the way in which a plantation master signified the act of taking possession of a slave by giving him or her a new name.

One of my relatives was a nineteenth century taxonomist, a botanist whose work was to classify and name innumerable African and European plants and flowers. He was doing what Adam did - appropriating the natural world by branding its hide with a word. That was a century ago. Science has moved on since then and now acknowledges, for example, the inevitable way in which study modifies and changes the object of research. Has religious thought moved forward to a point at which it fully acknowledges the integrity of creation in all its forms? Or is it holding on to the old concept of enlightened human stewardship for birds, fish and plankton, for landscape and seascape, mountain and steppe, shore and seabed, animate or inanimate?

Some of the most devoted environmentalists, always ready to criticise human intervention in the natural world, only understand the duty to care for other species in terms of a categorical difference between the children of Adam and the creatures allegedly placed in their care. Accepting the human race as an integral part of the natural world should not, however, demolish the imperative to care for life-forms and landscape.

Until we can get some consensus about the nature of human responsibility, we will find it hard to agree about quite concrete things. Is it, for example, only Man that can pollute? Cannot birds, poison-bearing insects or invasive comb-jellies pollute the environment just as human beings can? The gigantic process of pollution which rendered most of the Black Sea anoxic, poisoned with hydrogen sulphide and almost lifeless to within a few hundred metres of its surface, was, after all, caused not by Man but by the great rivers pouring out their organic debris during the ten millennia or so of the Holocene. Is it right to combat the damage done by the *Mnemiopsis* plague to the Black Sea by introducing alien fish species which will eat the jelly, knowing that by doing so we will be helping to construct a new and different ecosystem? Or is that an act of gross interventionism and arrogance and should we instead restrict ourselves to cleaning up rivers, reducing coastal sewage pollution and suppressing overfishing?

When we look at the Crimean landscape, which everyone should care for and nobody can possess, we should also remember that these Black Sea shores are all homes which have been lost by people who loved them, used them and cared for them. Millions of men, women and children were driven away from them by force and died far away, remembering them only in dreams and then in myths: the Pontic Greeks from Anatolia, the Meshketian Turks, the German colonists of south Russia and Ukraine, the Greeks of south Russia and the Caucasus, the Crimean Tatars, the Russians who fled through Novorossiysk and Sevastopol and Odessa to escape the Bolsheviks, the Mingrelians and Georgians who now live in terrible misery in Batumi and other Georgian cities after their flight a few years ago from Abkhazia. Far in the past, the Scythians and Sarmatians, the Crimean Goths and the Byzantine Greeks, were all expelled in turn from these coasts.

The Black Sea shores are now a holiday destination and although it should not be forgotten that mass tourism is an extractive industry, eventually wearing away and destroying the resource it exploits, weary people from northern cities are entitled to enjoy these shores and to fall in love with them. At the same time I hope that visitors will not quite forget the ghosts of those for whom these places were home, whose world was taken away from them even though they had done it no harm. This is a coast and a sea which deserve reverence:

> The Black Sea has a personality which is not caught by some adjective like 'unpredictable' or phrase like 'friendly to strangers' and which - because it is not made up of traits or epithets but of the interplay of circumstances - cannot be described in detail at all. These circumstances, adding up to an identity, include fish and water, winds and grass, cliffs and forests, migrating birds and human beings. This is not just a place, but a pattern of relationships which could not have been the same in any other place (ibid).

Beauty and the Divine in Nature

Bishop Basil of Sergievo

According to the sixth century theologian Dionysius the Areopagite, the fundamental name of God is 'good'. This essential good, by the very fact of its existence, extends goodness into all things (DN 4 693B). For Dionysius, therefore, what exists is good, and what is good is - among other things - beautiful.

The work of Dionysius has found an enduring place among the fathers of the Eastern Church and has deeply influenced the development of theology in the Christian West. Dionysius is concerned to give us a picture of the universe in which God is the source of all that is. He is the Creator, yet is radically different from his creation. He is the source of all things, whether visible or invisible. He is Being; he is Wisdom; he is Mind; he is Truth; he is Power; he is Beauty; he is Righteousness; he is Word; he is Faith; he is Redemption; he is Salvation. All that is comes from Him and all that is, is good.

In the world of Dionysius there are nine 'orders' of angels between God and the world in which we live. The lower order longs for God and thus longs to grasp and to understand that which is above it. In this it is positively helped by the higher order, whose task it is to purify, illumine and perfect the order below it. Thus in the Dionysian world there is a constant movement from above, through which the divine Goodness is mediated to the visible creation, and a returning movement from below, whereby the visible creation, through love (*eros*) returns towards God and in the course of this, is purified, enlightened and perfected. Dionysius, in his own way, and by the use of thought forms with which he is familiar, is expressing certain fundamental truths about the limitations of our understanding of the world.

Yet how are we to understand this hierarchy of 'intelligent and intelligible beings' who mediate for us the Goodness, the Beauty, and the very Being of God?

The Dionysian world can be compared with two aspects of modern thought: mathematics and the world of quantum physics. Bertrand Russell's *Principia Mathematica* had developed a generalised and all-embracing mathematical formulism that was intended to enable us to describe the real world in mathematical symbols and thus grasp reality in its entirety, but Kurt Goedel, an Austrian mathematician active in the 1920s and 30s, was able to show that logical thought is subject to certain unavoidable limitations. Specifically, he was able to demonstrate that any formal system rich enough to include mathematics is necessarily incomplete. This unsatisfactory situation:

> can be eliminated for any particular system by shifting it onto a wider system of axioms, within which we may be able to prove the consistency

> of the original system. But any such proof will still remain uncertain, in the sense that the consistency of the wider system will always remain undecidable.[1]

Part of what we find in Dionysius is an adumbration, in his own terms, of this mathematical truth. Each angelic order is able to comprehend and guide the order below it, but no one of them can comprehend or grasp the order above. The truth of any particular order is therefore contained not in itself, but in the order above. Though all the angelic orders are reflections of the truth, they are incapable, at their own level, of reflecting the whole of the truth, of *being* the truth, and in this they resemble axiomatic systems that have come to life, 'minds' (*noes*), which are structured in their relationship to each other much like the axiomatic systems of which Goedel speaks. In the Dionysian world an infinite regress is prevented only by God, the Creator of the angelic hierarchies, who is beyond and above all comprehension of what is created.

In the 1940s, A. Tarski went on to develop Goedel's arguments further, showing that if a factual assertion is made by a sentence *p*, then (in that particular axiomatic system) '*p* is true' is not a sentence. The effect of this would seem to be that the statement 'God is good' - statement *p* - cannot formally be shown to be true within the language in which it was originally made.

I suspect that in Dionysius we have an adumbration of Tarski as well. Dionysius accepts that the only way to speak the truth about God in the only language he has is to both affirm and deny at the same time. If you wish to express the truth that 'God is good', you will actually come closer to the truth if you assert at the same time that 'God is not good', the statement that 'God is not good' contradicting on a linguistic level the statement 'God is good' and thereby offering a chance to say something true about God within the language used, while on an ontological level it indicates at the same time that God is beyond all mere created goodness and therefore beyond comprehension of created beings and beyond all human language.

If we are to avoid an infinite regression of the type required by Tarski, we shall have to accept that our language must contain statements which are mutually contradictory. If we wish to understand God as he is, then we must transcend the affirmations we make, contradicting ourselves in order to grasp, in a way which ordinary reason cannot give us, the mystery of the Divine.

Dionysius speaks in this way of the world as he understands it:

> We cannot know God in his nature, since this is unknowable and beyond the reach of mind or reason. But we know him from the arrangement of everything, because everything is, in a sense, projected out of him, and this order possesses certain images and semblances of his divine

> paradigms. We therefore approach that which is beyond all as far as our capacities allow us and we pass by way of the denial and the transcendence of all things and by way of the cause of all things. God is therefore known in all things and as distinct from all things (DN 7.3, 869C-872A).

The divine paradigms of which Dionysius speaks are the source of all beauty in this world. They are in God, and thus not directly accessible to us, nor even to the seraphim. Yet Dionysius knows they must exist. There is an ontological gulf between them and creation which is bridged by participation, but the relationship is one of type and archetype, in which the original leaves its mark on the material upon which it is impressed without at the same time leaving behind any trace of its being.

The Dionysian world-view can also be compared with the work of David Bohm, long-time professor of physics at Birkbeck College, London, whose last work, *The Undivided Universe,* written with Basil Hiley, appeared posthumously in 1993 (Routledge, London and New York, 1995). The subtitle of this work is 'An Ontological Interpretation of Quantum Theory' and the book carries forward previous work in which Bohm sought to give a 'clear, intuitive understanding of the theory's meaning and a coherent cosmology without having to assume an outside observer'.

One of the difficulties theologians have in making use of modern quantum physics has been its generally accepted status as an epistemology and not an ontology. The position of Bohr and Heisenberg, that quantum theory is better understood as being concerned only with our knowledge of reality, and in particular with how to predict and control the behaviour of this reality, leads to the scientist, qua scientist, being content to know how his instruments will function. It is enough to systemise their readings: it is not necessary to give an intuitively satisfactory account of the physical processes that cause these readings to appear.

However, for most of us, who are accustomed to working with what we think of as the real world, of which we have an intuitive grasp, the non-intuitive character of quantum theory makes it very difficult to fit into our scheme of things. For this reason alone Bohm's work should be interesting to the theologian but in addition Bohm's position deliberately deals with the universe as a whole. It is not a 'theory of everything', but he does insist that in any understanding at any level *everything* must be taken into account.

Bohm also rejects the conventional view that the wave function, by itself, yields the most complete possible description of quantum reality - and therefore of all reality. He concludes that local 'hidden variables' are possible, bound up with the existence of the measuring device itself, and proceeds to rewrite the formulae accordingly. He then seeks to show that certain existing experimental results, hitherto puzzling, can be dealt with satisfactorily on the basis of these new formulations.

Finally, Bohm seeks to integrate the phenomenon of 'non-locality' into his conceptual system. Non-locality is the puzzling feature of the quantum world whereby particles may be strongly connected even where they are far apart. What happens to one affects what happens to the other even though they would appear to be unconnected. Non-locality would seem to contradict the theory of special relativity, which is generally held to be as fundamental as quantum theory itself. Bohm proposes to understand the theory of relativity as applicable and valid for all *statistical* quantum results, as well as for the large-scale manifest world. In the 'hinterland' however, in the world of the individual quantum process, non-locality - in both space and time - is possible.

The world of David Bohm is therefore a layered world. Behind the Newtonian world in which we spend our ordinary lives, there lies the world of quantum mechanics, where special relativity holds sway and the wave function fully describes the observable quantum effect. Behind the quantum layer lies another layer where, in the individual quantum process, finer, more subtle relationships can, with difficulty, be observed, some of them non-local and involving information travelling faster than the speed of light, a limit belonging to the conventional quantum world.

How do these worlds relate to one another? The first thing that must be said is that there is only *one* world. The quantum world is a limited perception of that one world, and the Newtonian, everyday world in which we live and act is an even more limited perception of that one world.

Bohm suggests that we must greatly modify the way we look at things. For example, wherever I happen to be in a room, the order and structure of the whole room has been, as Bohm says, 'enfolded' into a space so small that it can pass through the pupil of my eye. When this order is 'unfolded' on the retina and in the brain, it somehow gives rise to an understanding of the order to the whole room. Similarly, he says, the order of the whole universe is enfolded into each region and may be picked up by various instruments such as telescopes from which it is ultimately unfolded into a conscious awareness of this order. The whole is present in each part.

The world of David Bohm is thus an undivided and completely interpenetrating whole which can be understood and perceived at different levels; the end result of Bohm's approach is a universe which is made up of an 'order of ever increasing orders of subtlety in the implicate order'. Each of these orders would seem to be capable of embracing the order below it, but would not be capable of being 'grasped' by the lower. Something in the higher order would always be incapable of being fully expressed in terms of the lower.

This structure resembles closely that outlined, in very different language, by Dionysius in the sixth century. Dionysius was able to speak, in his day, using a language about the real world that was common currency among educated people in his time. Theologians, today, find this very hard to do. We very much need to find a contemporary language with which to speak about the world as we know it through science and then relate this to our spiritual concerns and to God. Our discourse must revolve around three poles: God, man and nature. If we cannot speak convincingly about nature, how are we going to speak convincingly about man who is part of nature, and about God?

An epistemological understanding may - and does - serve very well if what we seek is to predict and ultimately control the behaviour of the world; but we humans also need to commune with the world, to live with it at the level of contemplation. We can do this quite easily at the level of the explicate, perceptual world, but for the integration of our whole being, we will also need to be able to do this at the level of the mind. Dionysius was able to do this, and has enabled many others to do so as well. A modern Dionysius, to produce the same effect, will have to possess a language about the material world something like that of David Bohm, a language that opens out easily into the world of the mind and of the spirit.

For Dionysius, perceptible beauty is but a dim reflection of the unutterable Beauty of the Creator, and serves - or should serve - to point our minds and our hearts to its source. The beauty we observe in the Black Sea coastal landscape is an objective property of nature, quite as real as colour or sound, and is accessible to and perceptible by every human being. For the monotheistic faiths, the ultimate source of this beauty is God - the Source of all that is.

Herein lies the attractiveness of any region of the globe where men have never been, and of any plant or animal that we can observe 'in nature' The ugliness that we see is - sadly - the work of men. The inanimate world and the world of plants and animals conforms - to use the language of Dionysius - to paradigms that express the will of God, divine paradigms we are unable to perceive directly, but whose mediated presence, in a world that is structured in depth, we can intuitively perceive.

Mankind alone - together with a portion of the lowest angelic powers - does not conform to the divine paradigm he was created to reflect and therefore does not conform to the image of God in him. That image is not confined to his conscience, or to his reason, or to his radical freedom. It is found in the whole of his being. Each individual human being is a hologram of the universe: everything that is 'out there' is 'in here' as well. Each of us is a microcosm of the whole. That is why we can experience plants and animals as our sisters and brothers - their existence is implicate in the deeper levels of our being. Conversely, our existence, the deep structure of our being, is implicate in them, though the world works in such a way that this receives only partial expression.

Thus our ecological task is to find ourselves in the universe - and to find the universe, as intellectually and spiritually perceived, in us. Our understanding will never reach the depths that are in us. However, we do not have to know everything before we begin to act. We have never let ignorance stop us in the past. The truth of our actions will depend on our conforming to the deep structure of our own nature, and thereby bringing our mode of being, our mode of behaviour, into conformity with the will of God, which is known to us in part, at least, through the world.

All religious traditions have ways of helping their members to do this, and we must use the resources of our traditions for a common goal, a common good.

Note

1. Polanyi, M., *Personal Knowledge*, p.259, Routledge & Kegan Paul.

Place, Culture and History as expressed in Landscape

Nicoloz Beroutchachvili

Detailed study of the landscape of the Black Sea region is still at an early stage, but the landscapes of the Caucasus have been comparatively well researched. There is now data on the physical-geographical situation of the Caucasian landscape, its climatic, geophysical, ecological and other purely scientific characteristics. Detailed maps of the region have been compiled and there are a number of monographs on the subject. Through computerisation we have also been able to carry out a number of experiments for determining the place of Caucasus in the global environment, its effect on the global environment and how changes in the global environment effect the landscapes of the Caucasus.

Crimea and Bulgaria are also well documented. Ukraine and Romania have been partially investigated and study of the Turkish landscape in the Black Sea region is just beginning.

In general, the landscapes of the Black Sea region can be divided into three basic types.

1. The damp and subtropical plains of Kolhida have a relatively warm winter and are extremely humid. In Batumi, for example, there is an annual precipitation rate of 2,500 mm. Mtirala mountain, near Batumi has a precipitation rate of 4,000 mm per year, half that of the whole of Europe. Mtirala means 'Crying'.

2. The influence of the Mediterranean comes into effect west of Sinope where there is a typical Mediterranean landscape, a dry climate, and rainfall which is usually confined to the winter months. Sinope itself, Novorossiysk and Yalta enjoy a sub-Mediterranean climate.

3. Finally there is the steppe landscape of the Black Sea and Azov Sea coastal territories. These landscapes stretch in a wide line and turn into the so-called Balkanian landscape near Bulgaria. These regions have a temperate climate and are covered mostly with oak tree forests.

This is also a region of remarkable ethno-linguisitic diversity. Representatives of Caucasian, Slavonic, Roman and Tirkic linguistic groups live in the region and there are parts of the Caucasus where 33 different languages are spoken. Each of these countries has developed in its own way, with its own traditions, but at the same time they also share some common features. Christianity, for example, is prevalent in all the countries of the region.

However, when we talk about the culture of a country or of a region, we do not mean just its language and religion. There are also a number of common indicators related to the traditions and to the general situation within the region. Just as man, or an ethnic group can affect the shape of the landscape, so the reverse is also true - the landscape itself affects the human being. Everything is interconnected.

Every nation and every culture also has its own way of perceiving and of exploiting landscape. This can be easily seen in satellite photographs which show the borders of the countries and the borders between the different types of the land use.

Here in the Black Sea basin, two cultures - southern and northern - come together and their perceptions of the Black Sea are quite different. For Turkish people it is Kara Deniz - Black Sea. The Greeks, inhabitants of a warmer country, called it the 'inhospitable sea', whereas to northern people it is a warm sea. There is a Russian song in which the words: 'The most blue in the world, my black sea' are sung with great love.

These completely different approaches have themselves altered over the centuries and the history of the development of the Black Sea coastal territories reflects this.

For example, the western coast of the Black Sea was once considered the most unfortunate place on earth, rife as it was with malaria. After the so-called December Rebellion in Russia in 1825, some of the Decembrists were exiled to Gagra in north-western Georgia and some to Siberia. The Decembrists who had been exiled to Gagra appealed to the Tsar to be moved to Siberia because of the unacceptable conditions in the region.

At the end of the nineteenth century this perception changed. The Russian Empire needed resorts and so a Russian Riviera was created along the Black Sea coastline with some wonderful castles being constructed during that time. Later the Black Sea became the most popular place for tourists of the Soviet Union. Now the area is on the international tourist map and large-scale constructional work has been carried out to meet visitors' needs.

During the period of the Iron Curtain, significant work was carried out in Kolhida because there was no possibility of importing subtropical agricultural crops. Marshes were drained and intensive cultivation of subtropical agriculture was initiated. As a result of these human activities and the need for agricultural development, the landscape was moved ten degrees to the south

The north-western part of Turkey, being far from Istanbul, has remained undeveloped and the difference between the landscapes dividing Georgia from Turkey around the borders of Sarpi are so big that one appears to be moving from one continent to another. On the Georgian side there is subtropical agriculture, exotic flora , palm trees and bananas; on the other side, in Turkey, are wide expanses of untouched natural landscape, covered in parts by industrial waste. The shape of the borders has changed yet again more recently with intensive cultivation of tea plantations in Turkey, while in Georgia tea plantations are being destroyed because they are considered unprofitable. Consequently the borders have shifted to the north.

The current economic crisis in the region has had negative consequences on the landscape, such as the cutting down of forests; at the same time there have been some positive effects, because it has meant relief from concentrated activity. My parents were artists and as a child I spent a lot of time with them while they painted the Georgian landscape. They often argued about how the beauty and harmony around us came to be. My mother, as a true Christian, believed it was a gift of God. My father had his own point of view. I believe that it was this contact with nature that influenced my decision to become a geographer and to study landscape in depth. Georgia is a very beautiful country. It is also a developing one and it is essential that in the course of development we try to avoid creating ugly landscapes that will be difficult to reverse in later years.

Tourism: Enjoying without Destroying

Terry De Lacy

It is my belief that tourism has the capacity to be a dynamic engine, capable of driving individual and community action and change towards environmental conservation.

Our environment, our ecosystems, our natural areas have immense value in all sorts of ways: spiritual value, intrinsic value, ecosystem value, natural resource value. The majority of the value, however, is non-market value. Environmental goods and services are not generally traded in the market place. Our dominant market economy undervalues the environment and as a consequence it is degraded. We behave as we do for many reasons, but in our global, consumer-based market economy, our financial self-interest - whether individual, community or national - is one major determinant of why we do things. One of the challenges in conservation is to harness the values inherent in the environment and to incorporate them into our everyday economic activity, making it in our individual, community and national financial self-interest to conserve and protect the environment. Tourism is potentially one powerful means of doing this, by putting market value into environmental value. Large parts of our ecosystems and natural resources are public goods; no individual can exclusively own them. It is precisely these public goods, beaches, seas, landscapes, forests and wildlife, upon which tourism is built.

Tourism is certainly a major engine of our economy. It is the world's largest industry and generator of quality jobs encompassing transport, accommodation, catering, recreation and services for travellers. Look at the following statistics:

- Gross output 1997: US$ 3.8 trillion (world-wide)
- Gross output 2007 (predicted): US$ 7.1 trillion
- Employment 1997: 262 million jobs
- Employment by 2007 (predicted): 383 million jobs
- World GDP 1997: 10.7%
- World GDP by 2007 (predicted): 10.9 %
- 42.3% growth globally predicted over the next decade.

(Source: WTTC 1997.)

How then can we ensure that this large economic engine works for conservation? How can we achieve sustainable tourism?

If we consider tourism as a land use, compared with other industries it is potentially a non or low consumptive user of our natural areas and resources. Take for example a pristine forest landscape. How can we extract the value or use the forest to increase the welfare of our community?

- We can log it for timber or paper;
- We can cut the forest and grow wheat;
- We can clear the forest and establish residential or industrial development;
- We can dam and flood it, harnessing water for irrigation and hydropower;
- We can put a board walk through the canopy to support a tourist industry.

Using it for tourism means that we have low environmental impact, provide for local development, including often much needed employment, gain community and political support for conservation, educate the user (and employees) about the importance of forest ecosystems and potentially change the way they will act in the future regarding conservation.

Secondly, how can we ensure that tourism development is indeed sustainable, because not all of it is.

Tourism will be ecologically sustainable if it:

- does not use non-renewable resources faster than renewable substitutes can be found for them;
- does not use renewable resources faster than they can be replenished;
- minimises operational energy consumption;
- does not release pollutants faster than the biosphere can process them into a harmless state;
- has low impact on biodiversity and ecological systems and processes;
- maintains a full range of recreational, educational and cultural opportunities for the present generation and future generations;
- benefits local communities and the region socially and economically;
- does not affect the capacity of other sectors of the economy to achieve ecological sustainability.

To be truly sustainable it must of course be competitive in the market place and turn a profit over the long term.

There are many policy instruments that can be used to help the drive towards sustainability, including regulation, planning, industry codes of management practice, best management practice, and quality assurance accreditation.

A number of market forces encourage sustainable tourism ventures. These include:

- increased consumer resistance to degraded environments;
- evidence that sound environmental practices have long-term economic benefits;
- financial rewards and incentives for good environmental practices;
- fines for bad environmental practices;
- the likelihood of media exposure for bad practice developments;
- the growing demand for sensitive and innovative designs for developments in fragile environments.

A recent survey, conducted across the United Kingdom and involving face-to-face interviews with 2,048 people aged 15 years and over, provides some interesting insights into people's behaviour and their attitudes to tourism. Interview data was statistically weighted to provide an overall population profile. The results show that a majority, 52%, of people believe that when purchasing holidays or arranging business travel it is important to do business with travel and tourism companies that take environmental issues into account. Around five in ten people said they would be willing to pay more for their holidays or business trips to ensure the tour operator, transport company, hotel, tourist attraction, caterer, and retailer were committed to environmental protection. On average, they would be prepared to pay between AUS$10-$16 extra for each component (Mori/Green Globe 1995).

A survey of members of the Tourism Council Australia (TCA), the principle Australian tourism body which represents 30,000 tourism enterprises or organisations, found that they believe that of the nine most important attributes of their tourism business, seven were directly related to the environment.

Self-regulation is an important tool in achieving sustainable tourism development. The approach used by the Tourism Council Australia to establishing a sustainable code of conduct has incorporated the establishment of a series of sustainable tourism principles, a programme of best practice environmental management and monitoring of sector specific outcomes such as energy conservation, water conservation, waste management, wastewater treatment, environmental stewardship and community/cultural input.

The TCA's sustainable tourism principles include:

- using resources that are sustainable

- reducing over-consumption and waste
- maintaining diversity
- integrating tourism into planning
- supporting local economies
- involving local and indigenous communities
- consulting stakeholders and the public
- training staff
- marketing tourism responsibly
- undertaking research.

The Code of Sustainable Management is voluntary and open to all TCA members. It is directional, non-prescriptive and self-regulatory so as to encourage participation.

Signatories to the Code will have the following key obligations:

- to develop, implement and monitor sustainable tourism practices;
- to continually improve operations;
- to consult and communicate environmental performance to key stakeholders;
- to give priority to service providers and suppliers who show a commitment to sustainable business practices;
- to participate in the Environmental Category of the TCA state tourism awards; and
- to participate in environmental reporting activities.

As an example of best management practice, let us look at sustainable tourism development in a typical coastal zone, for example, the Kingfisher Bay Resort on Fraser Island off the coast of Queensland. There are five stages:

Stage 1: Feasibility

Important steps in establishing the feasibility of a proposal include:

- Developing a profile of the proposed venture, including intended market and competition, facilities and services, atmosphere and style, the quality and capacity of the operation, strengths and weaknesses of related developments, and the type of site that would be suitable;
- Consultation with relevant authorities prior to the application stage;

- Investigation of potential sites;
- Assessment of the proposal's ecological sustainability, to ensure the development operates within the capacity of the environment;
- Determining the environmental acceptability of the proposed development;
- Assessment of the likely economic viability of the proposed development.

Stage 2: Planning and design

- A responsible design for development is based on the needs of the site's environment and its local communities.
- Higher initial costs will be offset by the venture's long-term viability (ecologically sustainable developments are valued highly both in terms of their reputation and hence occupancy rates and resale value).
- An Environmental Management System identifies environmental targets and objectives and how it is proposed to achieve them; it also helps clarify the environmental issues faced, the long-term environmental objectives and the short and long-term social and economic benefits of achieving them.

Stage 3: Assessment and approval

Stage 4: Construction

The goal of this stage is to develop the tourism site with minimal environmental impact and maximum community support. Strategies to minimise or avoid construction impacts include:

- Measures to maximise recycling and reuse, for example, using cleared vegetation for mulch;
- Controls to prevent littering, vegetation damage, introduction of non-indigenous species, fires and firearm use;
- Measures to minimise soil erosion and to deal with wastes generated on site;
- Briefing/training for all contractors/employees to ensure standards and procedures are understood and applied.

Stage 5: Operation and management

- The goal of this ongoing stage is to review, ensure continuing implementation and evaluate the effectiveness of the environmental management system.
- Major activities will include repairs and maintenance, staff training to enhance

awareness of their responsibilities, visitor education, ongoing community contact and involvement and effective monitoring, evaluation and adaptation of the environmental management system.

As we struggle with our environmental issues - whether problem identification, description, cause, or more importantly solution, be it global, national, local, total or incremental - we need to engage those who work in other disciplines and professions, and through them the businesses, industries, departments and communities where they work and live.

Some people feel that recently the environment has declined as an issue and it is true that specialist programmes in environmental science or studies have declined in our universities. On the surface this seems worrying, but if we look a bit more deeply we will find economic degrees with compulsory studies in the environment; engineering degrees with majors on environment engineering or waste engineering; agricultural programmes embedded in natural resources degrees with a heavy concentration on sustainable farming; the mining industry showing significant environmental concern. In the faculty of Natural Resources and Agriculture at the University of Queensland our largest external source of research funds is for mine site rehabilitation.

In other words, the environment has become mainstream - it is becoming internalised in all our professions, disciplines and industries. As conservationists I believe that we put too much faith and responsibility in governments. Instead we need to engage directly with all the processes and institutions of development, whether they be local cooperatives or the myriad of business enterprises that make up our complex societies. Business enterprises, whether small, medium or global multinationals, are made up of people, and we need to persuade them, shame them, research with them, develop policy with them, and show them how better environmental management practices can indeed improve their bottom line.

Chapter 7

Industrial Impact

Introduction

Charles N. Ehler

In 1971 the British author C.P. Snow wrote cogently about what he called the 'two faces' of technology, benign and threatening. 'All through history it has brought blessings and curses. It was true when men first made primitive tools and clambered out into the open savannah: one of the earliest uses of those tools seems to have been made for homicide. It was true of the discovery of agriculture, which transformed social living but also made some sort of organised armies practicable. It was true of the first industrial revolution. Perhaps the sharpest example of this two-faced nature of technology is the effect of medicine. It has reduced infantile mortality, even in the poorest countries... Yet it has led us straight into the flood of population which is the greatest danger of the next fifty years.'The unexpected environmental consequences of pollutant discharges from industrial sources, the cost of pollution control (estimated to be about US$125 billion annually or over 2% of Gross Domestic Product for the United States alone), and continuing population growth and technological development are causing engineers to rethink the basic concepts of engineering design, as they relate to environmentally compatible technologies. There is a growing trend toward waste reduction, or pollution prevention, stressing total system design (as opposed to old add-on devices or end-of-pipe treatment) to reduce or avoid waste formation in industrial systems.

The focus of this chapter on the impact of industrial activities is marine navigation. More than 1.7 billion tons of crude oil are transported annually by ship from producing and refining countries to consuming countries. Oktay Ekinci describes the role of international marine commerce in the historical development of the Black Sea region and the threat of increased numbers of ships that will transport Caspian Sea and other Asian oil to its markets, particularly through the Bosphorus Strait.

John Lyras reviews the international legal framework that regulates the discharges of oil from both routine shipboard operations, for example, de-ballasting and tank-cleaning, to accidental spills. While marine transportation accounts for a relatively small amount of total oil pollution in the sea on a global basis, when major marine accidents occur they usually result in spectacular, short-term impacts, for example the Exxon Valdez oil spill in Alaska. Lyras concludes that the international regime has reduced significantly both operational and accidental discharges of oil into the marine environment and that voluntary shipping associations such as HELMEPA (Greece) and TURMEPA (Turkey), formed to educate the maritime industry about its environmental impacts, show promise for maintaining this trend.

Professor Patricia Birnie discusses the legal order of the seas that has evolved from the ‘freedom of the sea’ of Grotius (1609) to the obligations of all nations ‘to preserve and protect the marine environment from all sources of pollution’ under the Law of the Sea (UNCLOS 1982). She points out that it is only quite recently ‘that moving toward management of the seas as an integrated whole, rather than through a series of *ad hoc* regimes, has been seen to be desirable and *necessary*’.

Finally, a controversial proposal by Dr. Richard Harbison to remove an invasive species, the gelatinous ctenophore *Mnemiopsis leidyi*, that has been carried to the Black Sea from the Americas in the ballast waters of ships, is described in the last paper. The introduction of the ctenophore has dramatically altered the ecosystem of the Black Sea, including the decline of the anchovy fishery. Harbison would introduce another foreign species, the butterfish *Peprilus triacanthus*, to reduce the population of the ctenophore - an action that itself poses significant risks to the Black Sea ecosystem and one which provoked criticism from several local scientists in the symposium itself.

International and national actions, such as the Oil Pollution Act of 1990 in the USA, and subsequent rules and legislation, substantially increased the economic cost of marine pollution incidents to polluters and provided other measures intended to enhance marine safety and environmental protection. The changes have produced substantial incentives for shipping companies to reduce operational risks through improved ship construction, navigation procedures and equipment, and professional development. However, there is still no systematic programme to monitor shipping, economic, and safety performance trends or their relationship to marine safety laws and regulations or other options for improving marine safety. These monitoring and assessment programmes are essential for the maritime industry to continue to move in the right direction.

The Environmental Challenges of Coastal Cities

Oktay Ekinci

Most of the coastal cities of the Black Sea were built in the colonial period during the sixth and seventh centuries BC, taking shape as piers of world trade on the stage of history. When the sailors were building up these colonial cities, they were not building an ordinary coastal zone but the first port buildings and pier settlements in places rich in natural resources that offered people a living environment. Today, when we take a look at those same coastal cities, it seems that the world marine trade that presented those cities to humanity and civilisation 2,700 years ago now wants to take them back, perhaps not destroying them completely, but destroying them by polluting their life resources.

The ships that set sail in the seventh century BC from Phokaia (Foça) on the Aegean coasts of Anatolia, considered an ancient city even then, built up a city on the French coast (today's Marseilles) after sailing the Mediterranean from east to west. As a result of the Phokaians' activities, Marseilles became the centre of port trade for 26 centuries.

In the same century, perhaps in the same years, Byzas, the King of Megara, and his commanders set sail from the Greek peninsula to the east, and after passing the Dardanelles and the Sea of Marmara, established the Byzantian city in Sarayburnu which is today the historical centre of Istanbul. During the passing centuries, this city that has become Istanbul grew to embrace the Bosphorus and became the centre of the Eastern Rome, Byzantium and Ottoman states, the most famous world marine trade port in the Europe-Asia-Africa triangle.

In the same ages, the sailors that set sail to the Aegean Sea from Miletos (Milet), in the south of Izmir, changed their course to the north and sailed to the Black Sea and built up Kotyora city port near Ordu, the ancient Amisos city which is the place of origin of today's Samsun. Many cities on the coasts of the Mediterranean, Aegean and the Black Sea were created like this.

The environmental problems faced by all these coastal cities today are largely the results of the intensity of usage of the seas by the international marine trade. Almost all the countries in the world now use the port cities for shipping and freighting. Because of their commercial interests none are willing to take serious precautions to prevent the pollution of the environment of sea and city that their activities cause. In spite of providing an international service, these cities are deprived of international protection and supervision.

One example of this universal insensitivity is the intensified scale of the passage of oil tankers through the Bosphorus. International companies that market petroleum from the Caucasus, Khazar and Asia to every corner of the world use tankers as the basis for the transfer of oil from the Black Sea to the Mediterranean and there is no sign of any restriction of their use.

The Russian city of Novorossiysk is already threatened. When the Khazar and Kazakh petroleum flows through the oil pipes to this beautiful coastal city, and the new giant oil terminals have been constructed, this port will become a nightmare. Then the giant tankers, which have a capacity of 150 to 200 thousand tons, will turn to Istanbul with their cargo of petroleum.

The Bosphorus has for years been under threat from oil tankers, even before the shipping of the Asian petroleum began. With a length of 31 kilometres, its narrowest part less than one kilometre wide, 12 route changes and intensive use by civilian sea traffic, it is the world's most risky sea passage. There were 444 accidents in the 40 years from 1951 to 1991; between1980 and 1996 there have been 174. In a very high percentage of these accidents, the colliding ships were loaded with fuel oil and explosive chemicals. As a result of 50 accidents, the city was either endangered by huge fires which caused irreparable sea pollution, or many historical buildings in the coastal zone were damaged, thereby damaging the cultural richness of the Bosphorus.

With the likelihood of Novorossiysk becoming a more intensively used petroleum port, the accidents and the risks threatening Istanbul will increase by five or six times and protecting the Bosphorus and Istanbul from this enormous danger will become more and more difficult. There must be a universal effort to ensure that the transportation of the Khazar-Caucasus and Asian petroleum is done without polluting the Black Sea and the Bosphorus. Istanbul cannot take the risk of throwing away its 2,700 years of rich cultural history in front of the petroleum tankers.

The Shipping World and Protection of the Sea

John Lyras

I often find it difficult to reconcile my position as an individual and citizen of this planet with my position as a businessman and company operative. Showing respect for the environment and combating pollution are undoubtedly the main challenges of our times on a scientific, political and ethical level. This I fully acknowledge as an individual. As a businessman I have to abide by the rules of the global market place and global competition if I am to remain in business.

The shipping industry was the first mode of transportation to acknowledge the need to address the environmental impact of its activities. As early as 1954 the international shipping community through the International Maritime Organisation (IMO, formerly IMCO) adopted the OILPOL Convention. This was subsequently replaced by the MARPOL 73/78 Convention, the International Convention for the Prevention of Pollution from Ships.

There are two ways in which ships can pollute the environment: as the result of normal operations and as the result of an accident.

Operational pollution principally involves oil - in various forms - entering the sea. Ship design has played an important part in contributing to this problem. Traditional tanker designs rely on sea water carried in cargo tanks to stabilise the vessel while sailing empty. Imperfect cleaning of cargo tanks combined with the carriage of oil and then water in the same tanks constituted a major problem in the 1960s and 1970s when the number and the size of vessels increased dramatically. Since then, crude oil washing methods and the use of segregated ballast tanks have reduced this problem dramatically and tar balls, a frequent sight on beaches and coasts ten to 15 years ago, are now a rarity.

The other form of marine pollution is one which follows an accident, either a collision, grounding, fire or explosion. In 1990 the United States National Academy of Sciences estimated that tanker and non-tanker accidents accounted for 121,000 tons of oil entering the sea per annum compared with 411,000 tons entering the sea from operational discharges. According to IMO's Group of Experts on the Scientific Aspects of Marine Pollution (GESAMP), normal sea operations appear to be responsible for approximately 70% of the total oil pollution originating from ships, compared with 21% due to accidents.

The efforts of the shipping industry and the development of navigational aids reduced the number of accidental spills dramatically in the period from 1976 to 1995. New tanker designs and operational measures have reduced the quantity of oil spilled in cases of accidents even further. According to statistics collected during this period by the International Tanker Owners Pollution Federation, for larger spills (over 700 tons of oil spilled), the number of incidents per annum has dropped from 25 to three, and for small spills (below 700 tons), from 68 to 20.

Operational pollution has been diminishing at the same time. According to the United States Academy of Sciences, the oil entering the sea from maritime transport fell from 2.13 million tons in 1973 to 0.411 million tons in 1990. This is 100,000 tons less every single year for the 16 year period covered by the statistics. In the same period the tanker fleet increased its capacity from 186 million dwt to 232 million dwt. This is an achievement. In 1973 there was operational pollution to the tune of 11.4 litres per dwt per annum. In 1990 the figure had gone down five times to 2.4 litres per dwt per annum, and continues to decrease.

In terms of volume, oil pollution is by far the greatest form of marine pollution caused by ships. It is not, however, the only one and oil pollution by hydrocarbons, especially crude oil, is essentially non-toxic in contrast to other forms of pollution and there is evidence that ecological balance is restored - or even enhanced - within a few years. Yet this type of pollution attracts enormous media attention. Other forms are caused by: chemicals; dangerous goods (in bulk and packaged form); garbage; sewage; ballast water and anti-fouling paints. For each one of these categories of pollution there is international legislation extant or in progress.

While shipping has been addressing the environmental concerns of marine pollution in a systematic way for which the industry and the IMO should be commended, there is no doubt that more can be done especially in terms of fuller and universal application of existing requirements, particularly the provision of adequate reception facilities for oil-contaminated water in ports. This matter is the responsibility of signatory states to the MARPOL Convention and has been of concern for 20 years. The developed world has little or no excuse for the continued lack of proper facilities, and while the recent involvement of the European Commission and the possibility of a useful European network that will serve both public concerns and ships at an affordable cost are welcome steps forward, a European network alone is not enough

Pollution of the sea can be prevented by other means. Vessel Traffic Systems - if properly operated - contribute to fewer vessel strandings and collisions. Modern navigation technology can provide superior assistance to bridge and pilots.

Pollution from ships also results from the fact that ships are powered by thermal engines which, like all internal combustion engines, produce emissions. IMO has been active in preparing specifications for cleaner air, targeting atmospheric pollution, and government representatives in IMO are currently completing the new Annex VI of MARPOL which deals with atmospheric pollution from ships. The far-reaching provisions of this Annex, which covers exhaust emissions, cargo vapour emissions, ozone damaging substances and noise, will, it is hoped, successfully address the concerns of those states which have been pressing for the designation of their regions as 'special areas' affected by air pollution from shipping. The other big area for potential improvements is atmospheric inputs. The main areas where improvements are expected following the application of the new Annex VI of MARPOL are in sulphur dioxide and nitrogen oxide contained in exhaust fumes. The former will come from improvements in fuel quality, the latter from improvements in engine designs. It remains to be seen if these improvements will be matched by the main protaganists of atmospheric pollution, road, air and rail transportation, and of equal importance, factories, power stations and oil-fired domestic heating installations. Improvements in one category of polluters can easily be neutralised or even offset by deteriorations in another category.

This is a formidable array of legislation, the application of which will soon be subject to checks by the Port State Control authorities, and one might legitimately ask if the problems of marine pollution are on their way to becoming resolved. The answer is, sadly, not, and that is because shipping is only one and in reality the least offensive of the contributors to environmental pollution. Shipping is responsible for 12% of marine pollution. Even if shipping were successful in totally eliminating its own contribution, the problem would remain unresolved as long as land-based discharges (44%), atmospheric inputs (33%) and dumping (10%) are not brought under control.

Land-based discharges are of particular importance to the Black Sea due to the practice of using the big rivers as sewers and recipients of all kinds of industrial and other toxic waste, including agricultural runoff made toxic by extensive use of chemical fertilisers. The reason this happens is that it can be done and remain undetected. Even when it is detected, the consequences are still negligible in most countries. The river then takes the problem away from the polluter and places it into the defenceless sea. Some elementary discipline on the part of riverside states is necessary in order to prevent it. However, whereas the problem of controlling individuals who pollute the rivers can be resolved, preventing state-owned or run enterprises and organisations from polluting is a good deal more difficult to achieve.

It is commonly thought that dumping at sea is a problem that originates from shipping. However, although ships are used for dumping, it is a governmental activity and should be dealt with as such. Dumping takes place on issue by a country of a licence against compensation to a party to dump a certain quantity of waste into the sea.

There are specific international arrangements in force today for the compensation of the victims of oil pollution following a ship accident. The two IMO Conventions, CLC (International Convention on Civil Liability from Oil Pollution Damages) and FUND

(International Convention on the Establishment of an International Fund for the Compensation for Oil Pollution Damage), provide compensation to victims via a complex system whereby the ship pays up first - up to a limit - and the cargo owners take up the balance. This unique system, which has been in force since the late 1970s and whose liability limits were enhanced by the 1992 protocol, has been running successfully for 18 years. It is disappointing to see that only one of the Black Sea states, the Russian Federation, is a signatory of the FUND Convention.

The International Convention on Liability from Hazardous and Noxious Substances (HNS) adopts the same principles in providing compensation to victims of ship generated pollution other than oil in the event of an accident. In view of the new sensitivity of the Black Sea states regarding marine pollution matters - so clearly exemplified by the Interparliamentary Conference on the Environmental Protection of the Black Sea in Istanbul in June 1997, and the International Conference on the Ecological Safety of the Black Sea in Gelendzik, Russia, in October 1996, I would strongly recommend that the Black Sea states re-examine the advantages of participation in the CLC/FUND and HNS Conventions. This is especially relevant after the withdrawal by the industry of its voluntary schemes, the TOVALOP and the CRISTAL which were set up 25 years ago to facilitate the compensation of oil pollution victims and the costs of oil spill clean-up until such time as the CLC and FUND Conventions assumed broad support.

Greek shipowners and crews have put together their own association, HELMEPA, the Hellenic Marine Environment Protection Association. This association deals mainly with environmental education for the industry's personnel. The success of this scheme has raised considerable interest in other parts of the world, and Turkey has now formed a similar association, TURMEPA. Such initiatives, based as they are on voluntary commitment, are as important, if not more so, than regulatory regimes and are to be wholeheartedly acknowledged and encouraged.

There is no doubt that there are still areas requiring attention and, unfortunately, accidents at sea cannot be totally eradicated. I genuinely believe, however, that international shipping is living up to its obligations to the public at large and that Greek shipping in particular can show remarkable progress in the areas of pollution prevention and response. IMO has been addressing the problems in such a comprehensive manner that very few industries can claim to offer comparative coverage.

However, unless the requirements for the control of pollution are universal, competitive distortions and inequities will always set in, fostering resistance and hindering progress. Though pollution itself has no ideological affiliations, it is a known fact that state produced or tolerated pollution is the hardest to eradicate. Pressure from environmental organisations, the scientific community and common citizens does have an effect, albeit limited, and that is why conferences such as this are vital.

Major Waterways: International Law and the Commons

Patricia Birnie

The terms 'international waterway' and 'the commons' are not clearly defined in either custom or treaty. *The Shorter Oxford Dictionary* offers as a definition in the case of the former, 'a route for travel or transport of water; a river or canal or a portion of the sea or lake viewed as a medium of transit; an opening for the passage or leaving of vessels'. It defines the latter only in terms of 'the commonality, the lower orders; provision for a community in common; the share due to each member'. 'Common,' however, is defined as 'belonging equally to more than one, belonging to all mankind alike ... belonging to the community; free to be used by everyone'. It has been left to international law to endeavour to refine and update these concepts to accord more clearly with contemporary environmental and developmental realities falling within its sphere.

In the intervening years, however, international law has been invoked to establish a variety of regimes for rivers, canals, lakes and straits, by means of both conclusion of treaties and development of new customary principles and concepts, which not only lay down the terms of use by bordering states and other users, but also aim to preserve and protect the marine environment. These culminated, in the case of the seas, in the United Nations Convention on the Law of the Sea (UNCLOS 1982), which aims to establish a 'legal order for the seas and oceans' to 'facilitate international communication, promote the peaceful uses of the seas and oceans, the equitable and efficient utilisation of their resources, the conservation of their living resources, and the study, protection and preservation of the marine environment' (Preamble). Achievement of these goals is recognised as contributing to 'a just and equitable international economic order which takes into account the interests and needs of mankind as a whole'. However, only the seabed and ocean floor and their subsoil beyond the limits of national jurisdiction and their resources are recognised as constituting 'the common heritage of mankind' (the Area), all rights to which are vested in mankind as a whole, on whose behalf an Authority instituted by the UNCLOS shall act on terms and conditions specified by it, as prescribed by the Convention and a modifying Agreement concluded in 1994. This Authority, however, is responsible for adopting measures to protect the marine environment from harmful effects which may arise from such activities in the Area.

Over recent years international law has subjected coastal waters to increasing national jurisdiction and control, while recognising that some areas (such as the high seas and the deep seabed) remain beyond national jurisdiction. In such areas, often loosely referred to as the 'global commons', states have certain common interests. In the case of the deep seabed, a doctrine of 'common heritage' has been applied. This resulted in the unique establishment by the UNCLOS in 1982 of an International Seabed Authority to administer it on mankind's behalf. No such 'Area' exists in the narrow confines of the almost enclosed Black Sea, however. In certain special cases international law has acknowledged the existence of 'common interest' in activities or resources that take place or are found wholly or partly within areas falling under national jurisdiction. Thus measures have been agreed aimed at protecting certain living resources, the atmosphere, biological diversity, or discharge of gases that might cause climate change. On the other hand, resources found and taken in the high seas beyond national jurisdiction have long been regarded as 'common property', which could be exploited and appropriated by citizens of all states on a 'first come first served' basis.

The concept of parts of the seas and oceans as some form of commons has been with us for a long time. It first came to the attention of lawyers during the great doctrinal debates started by Hugo de Groot (Grotius), a Dutch lawyer, in 1609. It has been evolving to adapt to different situations ever since, and still retains its usefulness in introducing ethical considerations into the international decision-making processes.

Ever since man developed the ability to sail far from home, disputes have inevitably arisen concerning the legal status of the waters navigated and the respective jurisdictional rights of coastal states and the states whose flag a vessel flies. Grotius instituted, in the phrase *mare liberum*, a standard that established awareness of international community interests, rather than exclusively national ones. He contended that states could not acquire title to areas of the seas by occupation, since such areas were *res communis omnium*, property that belonged to everyone.

Other states and jurists, however, opposed this view throughout the seventeenth and eighteenth centuries, supporting the concept of *mare clausum*, the ability of states to close areas of sea to foreigners, mainly foreign fishermen. By the late eighteenth century most national claims to vast sea areas had died away and by 1930, when the first Law of the Sea Conference was held at The Hague under the auspices of the League of Nations, one of its rapporteurs found it to be 'an unimpeachable axiom that every nation is free to travel to every other nation and trade with it'. Moreover, he regarded this as an axiom laid down by God himself through nature 'because not every place has the necessaries of life'. The sea was seen as 'God's highway'.

For nearly 400 years the principle that the seas are a global commons has served us well, since it requires us to evaluate, against the interests of the community as a whole, the propriety and legitimacy of every individual state's claim to diminish the commons in some way.

That said, however, even Grotius recognised that coastal states could exercise what he called *imperium* (sovereign rights) over a narrow belt of sea round their coasts in order to protect their interests in fishing and security, as well as international interests in freedom of navigation. Ensuring free navigation for other states in what became known as the territorial sea became the coastal state's responsibility. This evolved into the doctrine of 'innocent passage' - passage which did not threaten the peace, good order and security of the coastal state. This doctrine was not finally settled in its modern form until the conclusion of the UNCLOS 1982. By this time the threat of pollution by deliberate discharges or through accidents involving ships carrying oil, radioactive materials or other hazardous cargoes was causing concern to many coastal states. They challenged the 'innocence' of such passage and contended that passage of vessels posing environmental threats could be denied by coastal states.

The limit for this coastal belt was eventually settled at 12 nautical miles by the UNCLOS 1982. Various new types of zones were introduced and the powers states could exercise within them were refined and balanced. However, some straits used for international navigation fell wholly within the territorial sea of one or more of the bordering states that faced each other across them. A few states in this situation had, from the late nineteenth century onwards, negotiated a special treaty regime with other states concerned, including the flag states whose vessels were navigating such straits. This was the case in the series of treaties concerning the Bosphorus, the Sea of Marmara and the Dardanelles which culminated in the Montreux Convention of 1936 covering passage through these Turkish Straits. These treaties opened up international access to the Black Sea, which had been closed until 1774.

The UNCLOS 1982 preserves such existing treaties but establishes a new regime - a right of unimpeded 'transit passage' - through most other international straits. Bordering states cannot impede passage though they can regulate it in respect of the prevention, reduction and control of pollution. They can also designate sea lanes and prescribe traffic separation schemes for navigation where necessary to promote safe passage.

The UNCLOS also introduced a new jurisdictional zone, the 200 nautical mile Exclusive Economic Zone (EEZ). In this zone, coastal states are awarded 'sovereign rights' for purposes of living and non-living resource exploitation and 'jurisdiction' (the power to make and enforce laws and regulations) with regard to protection and preservation of the marine environment and marine scientific research. However, in order to protect the common interest in freedom of navigation, promotion of marine scientific research, preservation of the health of the oceans and other international concerns, these powers were made subject to other relevant provisions of the convention.

Acceptance of the 12 n.m. territorial sea and the 200 n.m. EEZ further eroded the area that Grotius had regarded as the *res communis*. To take account of this, the coastal states, in return for international acceptance of the extension of their jurisdiction, undertook to have regard to the rights of other states and act compatibly with the UNCLOS provisions, including those on preservation of the marine environment and conservation of its living resources.

During the 1960s and early 1970s the number of members of the United Nations and other international bodies rapidly increased as former colonies became independent and the UNCLOS III enabled new states to challenge some of the existing concepts of the law of the sea. Environmentally concerned states and non-governmental groups also seized the opportunity to press their interests. In 1972 the United Nations Stockholm Conference on the Human Environment (UNCHE) had adopted a Declaration of Principles and Action Plan and established the United Nations Environment Programme to promote them. These principles recognised the need to provide man with an environment of quality and also his responsibility to protect and improve it for present and future generations. They pronounced that toxic substances and other pollutants should not be discharged in quantities that exceeded the environment's (including the marine environment's) capacity to render them harmless and must be halted to ensure that serious or irretrievable damage is not inflicted on ecosystems (Principles 1 and 6 respectively). States were required to take all steps to prevent pollution from substances likely to cause hazards to human health, harm to living resources and marine life, damage amenities or to interfere with other legitimate uses of the sea (Principle 7). The right of states to exploit their own resources pursuant to their own environmental policies was recognised, but was coupled with the responsibility to ensure that activities within their jurisdiction or control did not cause damage to the environment of other states or of areas beyond the limits of national jurisdiction (Principle 21). All these principles, which protect both the common interest and the international commons itself, were incorporated into the UNCLOS 1982 in the part devoted to 'Preservation and Protection of the Marine Environment'(Part XII).

There has since then been an increasing use of non-binding means of augmenting or re-directing the goals and values of international society, such as declarations, resolutions, codes, charters, recommendations, guidelines and the like, and the promotion of generalised principles. The new United Nations members also introduced the concept that the areas of the deep seabed beyond national jurisdiction should be regarded as the 'common heritage of mankind', and persuaded the United Nations General Assembly to adopt a Declaration of 15 Principles proclaiming and giving effect to this approach in 1970. A key aspect of this concept is that the 'resources of the area' of the deep seabed shall be exploited only for peaceful purposes and for the benefit of mankind as a whole. This remains a key element of the UNCLOS 'package'.

The UNCLOS Part XII provisions state categorically that all states, flag states as well as coastal states, have the obligation to protect and preserve the marine environment from all sources of pollution. Moreover, states are required to cooperate and to harmonise their policies in this connection. They must protect and preserve rare or fragile ecosystems, as well as the habitats of depleted, threatened or endangered species and other forms of marine life. States must not transfer damage or hazards from one area to another or transform one type of pollution into another. They are also required to cooperate on a global or regional basis in formulating rules, standards, recommended practices and procedures consistent with the UNCLOS, taking into account characteristic regional features.

The high seas, though much reduced in area by EEZs and by expansion of the limits of other forms of zone, still remain, in the common good, open to all states for the exercise of the traditional freedoms, although not subject to the common heritage approach applied to the deep seabed. Any formerly implied unfettered freedom to dispose of pollutants in the high seas, however, is not recognised in the UNCLOS. The UNCLOS encourages states bordering enclosed and semi-enclosed seas 'to co-ordinate the implementation of their rights and duties with respect to the marine environment' and undertake joint programmes of scientific research (Part IX). However, it is only quite recently that moving towards management of the seas as an integrated whole, rather than through a series of *ad hoc* regimes, has been seen to be desirable and necessary.

Negotiating highly specific global agreements has become very complex. In 1992 the United Nations Conference on Environment and Development, held in Rio de Janeiro, adopted a new Declaration of Principles on Environment and Development and an Agenda for action to make it operational into the twenty-first century (Agenda 21). These introduced new responsibilities and reformulated some of the older ones laid down in the UNCHE Declaration of 1972 and promoted the concept of 'sustainable development'.

Agenda 21, which sets out the actions required to fulfil the Rio Declaration's aims, includes a chapter (Chapter 17) which addresses the 'Protection of the Oceans, All Kinds of Seas Including Enclosed and Semi-Enclosed Coastal Areas and the Protection, Rational Use and Development of their Living Resources'. Whilst this does not describe the seas as 'commons' it does propose new approaches which involve that analogy and acknowledges that: 'international law, as reflected in the UNCLOS, provides the international basis on which to pursue the protection and sustainable development of the marine and coastal environment and its resources.'

Many existing treaties protecting the seas and oceans have been re-negotiated to reflect these requirements.

From the non-binding processes, often referred to as 'soft law', and including UN, UNCHE and UNCED declarations and resolutions, have emerged such new concepts and principles as 'common heritage', 'common interest', 'common concern', inter- and intra-generational equity, the polluter and user pays principles, the precautionary principle, and environmental impact assessment. Although not well defined, they augment both long-established and recently crystallised principles of international law, such as equity, good faith (*bona fides*), good neighbourliness, cooperation and communication and have already had considerable impact in redirecting international maritime and environmental law.

Non-governmental organisations have played a major role in creating public awareness and lobbying to bring about these changes in international law. The religions also have much to contribute in bringing the international community to the realisation that the seas remain, as lawyers of the Grotian period contended, 'God's Highway'.

Controlling an Exotic Species in the Black Sea

Richard Harbison

The problem that I hope to help solve is one posed by an exotic species, the ctenophore, *Mnemiopsis leidyi*. An exotic species is an organism that has been moved by human activity into a region where it did not exist previously. In principal this definition seems simple but in practice it is often quite difficult. There are two problems. First, one must have an extensive knowledge of the biogeography of the species in question and know all the species that are native to the new region. Second, there is the fact that Nature also moves species around. In the case of the Black Sea, the primary natural mechanism that altered the composition of species there has been the periodic opening and closing of the Bosphorus. Relatively recently, about 20,000 years ago, a spectacular ecological disaster occurred - the Black Sea lost its connection with the Mediterranean.[1] This led to the extinction of all its marine plants and animals. Had people then had the technology that we have today, they might have attempted to keep the Bosphorus open, thereby preserving its marine life.[2]

After about 10,000 years as a lake, another ecological catastrophe occurred. The Bosphorus reopened, and the Black Lake was transformed once again into the Black Sea.[3] As the salty water poured in, large areas of land were poisoned. The Black Sea became much as we know it today about 3,000 years ago.[4] The Black Sea is now about half as salty as the Mediterranean, and the building of dams across the rivers that flow into it is helping to make it even saltier. Today, about 80% of the species in the Black Sea are immigrants from the Mediterranean. Even so, not all the Mediterranean invaders can survive in the Black Sea - in fact, most of them cannot. For example, there are about four times as many species of fish in the Mediterranean than there are in the Black Sea.[5]

Although the catastrophes caused by nature are the primary forces that created the present composition of the flora and fauna of the Black Sea, human beings are playing an increasing role in altering the list of species in the Black Sea. Shipping, which unintentionally brings in unwanted organisms on the hulls of ships and in their ballast water, is an important factor. So also are intentional introductions, such as aquaculture and introductions to improve the environment.

The way we live from day to day would be dramatically different without exotic species. Most of the foods on which the people of the world depend come from only a few areas of the Earth. For example, wheat comes from the Middle East, maize from Mexico, potatoes from Peru and rice from China. Most of our livestock comes from the Middle East and India. Any discussion of exotic species must recognise that exotic species have been a tremendous blessing to vast numbers of human beings. Indeed, the use of exotic species can be said, together with the prevention and control of disease, to have been one of the most important causes of the rapid increase in the number of human beings that has occurred over the past two centuries.[6] Because we depend so heavily on exotic species in agriculture, people's attitudes toward exotic species that live on the land are often very different to their attitudes toward exotic species that live in the sea. Most gardeners in the United States think little of planting ornamental plants from New Guinea or Asia, yet most ecologists in the United States regard the intentional introduction of exotic species into the sea as a bad thing. However, some countries, like Japan, consider the intentional introduction of exotic organisms to be a valuable way to improve and enlarge their food supply.

Determining whether an exotic species is good or bad is not easy. In the past, the main criterion was its effect on human activities. Bad exotic species included organisms that caused disease, or organisms that decreased agricultural yield. For the marine environment, species that decreased the abundance of commercially valuable fish were considered bad. Recently, however, with the increase in environmental awareness, the concept of a bad exotic species has become broader, expanding to include species that have negative effects on local animal and plant communities, even if they appear to have no effect on human activities. In practice it has become increasingly difficult to draw a clear line, so that some ecologists now consider exotic species as intrinsically bad. The inherent paradox in this position can be illustrated by the fact that these same ecologists may sit down to a meal of bread, potatoes, broccoli and chicken, all of which were originally found in different parts of the world.

Exotic species with negative effects on plant and animal communities can be considered a form of biological pollution. However, this biological pollution differs in one important way from other kinds of pollution - it is self-replicating. This means that the problem cannot be solved by simply stopping further introductions. Once an exotic species becomes established, it will continue to alter its new environment unless steps are taken to reduce or eliminate it.

Since the Black Sea is now in transition from a freshwater to a marine environment, it is an easy target for an exotic species, if that species can live there. One of the major reasons that exotic species can become very abundant is that they leave their diseases, parasites and predators behind when they move into the new environment and there is no control on their abundance except the availability of food and environmental conditions. The Black Sea has a very small species list, and appears to be particularly vulnerable to exotic species.

The most spectacular example of the devastating effect of an exotic species on the marine environment took place in the Black Sea when the ctenophore, *Mnemiopsis leidyi*, arrived in 1982, probably in the ballast water of ships moving cargo between the

Black Sea and the Americas. By the summer of 1988, it had become the most abundant organism in the Black Sea. As its numbers exploded, stocks of two commercially important fishes, the anchovy and the Azov Sea kilka, plummeted. It soon spread into the Mediterranean. Since its peak in 1988, its abundance has declined somewhat, although there is little data available to document this decline. Catches of anchovy have increased, although the significance of this is difficult to interpret, since there is practically no fishing effort from fleets of the former USSR. Catches of Azov Sea kilka remain low. There can be little question but that the introduction of this ctenophore into the Black Sea has drastically altered the structure of the Black Sea ecosystem.

The introduction of an exotic organism into an ecosystem is analogous to the introduction of a new disease into a human population. Just as with public health, the most cost-effective way to control the spread of an exotic species is with quarantine. Methods are being developed to control the spread of exotic species through ships' ballast water and protocols have been developed to control intentional introductions. However, the problem still remains - what do we do about an exotic species once it has become established? On land, pesticides are often used to protect crops, but we do not have this option in the sea. The problem in the sea is that the physiology of the pest is very similar to the physiology of the crop, so any pesticide that kills the one will probably kill the other. Also, even if a highly specific pesticide could be developed, its effects would be diluted, since pesticides do not stay in one place in water as they usually do on land.

Biological control is one option. This means introducing a disease, parasite or predator of the pest species. The place one usually looks to find such a control agent is the home range of the pest, since it is there that its enemies also evolved. I may have found a biological control agent that can reduce the harmful effects of *Mnemiopsis leidyi* on the Black Sea ecosystem. This animal is a predator, the commercially useful fish, *Peprilus triacanthus*. Commonly called the butterfish in the United States, this fish evolved in parts of America where *Mnemiopsis leidyi* lives and has evolved to feed on it and on other gelatinous animals. Besides ctenophores, it eats medusae and salps, which are also gelatinous. Since gelatinous animals are mostly made of water, it has to eat large amounts of prey. In fact, it eats its own body weight in ctenophores every hour.[7] In addition to its feeding being very specific, it can live in water that is almost fresh, just as the ctenophore can. This means that it can seek out and feed on the ctenophore wherever it occurs in the Black Sea - the ctenophore has no refuge from this predator. Whether it can in fact not only survive but reproduce in the Black Sea will require many more experiments in collaboration with scientists in Black Sea countries. Even if it does not eliminate the ctenophore, it will be converting the ctenophore into a commercially valuable fish, which can be used by people for food.

The intentional introduction of yet another exotic species into the Black Sea poses serious ethical questions. We must face the fact that when we introduce a new species, we alter the nature of the biological community. My suggestion is that we alter it in a way that is beneficial to humans and other organisms by converting ctenophore biomass into fish biomass. We do have to face the possibility that something could go terribly wrong and we have the responsibility to do everything in our power to avoid a negative outcome. However, as it is impossible to predict everything that could go wrong, we are faced with the choice of action or inaction.

Some might advocate inaction as the most prudent solution. Certainly it is the easiest. However, such a decision means that we have decided to let the ctenophore reproduce unchecked in the Black Sea, doing nothing to clean up the mess it is causing. Doing nothing seems to me to be the same as dumping oil into the sea and saying, 'Let nature clean it up.' For most of our history humankind has had this attitude. I think that it is now time for a change. We have to take action.

Notes

1. According to Degens and Ross (Degens, E.T,. Ross, D.A. (1972) Chronology of the Black Sea over the last 25,000 years, *Chemical Geology*10: pp.1-16), the transition to fresh water was evident at 25,000 BP (Before Present), and the Black Sea was truly fresh at 22,000 BP.

2. The idea of keeping the Bosphorus open to preserve the marine life in the Black Sea is not all that far-fetched. For example, the construction of a dam across the Strait of Gibraltar was recently proposed (Johnson R.G. (1997) Climate control requires a dam at the Strait of Gibraltar, *EOS* 78(27), pp.279-281).

3. The date of the reopening of the Bosphorus has been variously estimated as occurring as recently as at about 7000 BP (Glenn, C.R., Arthur, M.A., (1985) Sedimentary and geochemical indicators of productivity and oxygen contents in modern and ancient basins: the Holocene Black Sea as the "type" anoxic basin, *Chemical Geology* 48, pp. 325-354), and as long ago as at 9800 BP (Jones, G.A., Gagnon, A.R. (1994) Radiocarbon hronology of Black Sea sediments, *Deep-Sea Research 1,* 41(3) pp. 531-557).

4. Degens and Ross, op. cit.

5. Fischer, W., Bauchot, M.L., Schneider, M., eds. (1987) *Fiches FAO d'Identification des Espèces pour les Besoins de la Pêche. (Révision 1), Méditerranée et Mer Noire, Zone de Pêche 37, Volume II Vertébrés,* FAO, Rome; Svetovidov, A.H. (1964) *Fishes of the Black Sea,* Nauka Press, Moscow (in Russian).

6. It has been hypothesised that the increase in the population of Ireland between 1760 and 1840 from 1.5 million to 9 million occurred largely because of the introduction of the potato as a staple crop (Hobhouse, H. (1986) *Seeds of Change, Five Plants that Transformed Mankind,* Harper & Row, New York, pp. 205-206).

7. Harbison, G.R., Peharda, M., DeRobertis, A., ms. in preparation.

Chapter 8

The Implications of the Global Economy

Introduction

Daniel Amit

While religion and science can and must come together, neither of them affects the political process directly. And it is, of course, the political process, local, national and global, that must be changed to stop and reverse current trends which, in the tradition of this symposium, can be labelled 'Apocalyptic'.

One necessary bridge for linking religion and science on the one hand and the political/economic realm on the other is social science, primarily economics and political science. These disciplines are needed to assess the costs and benefits of politically expedient decisions and must weigh the costs of continuing along dominant ideological trails against information accumulated concerning the consequences. However, given that the tension between the practicalities of the real world and the inertia of intellectual tools may become very pronounced, the underlying assumptions of these disciplines must also undergo scrutiny.

In this chapter these issues are discussed by four voices. They all share the view that the situation is ominous and that the writing is on the wall for everyone to read. Yet the variety of approaches creates a mini-symphony. At ground level we have an unabashed, almost theory-free exposition by Edward Goldsmith which puts before us the direct consequences of 'globalisation', read simply as export-directed development. The conclusion is that this particular force necessarily undermines any possibility of the development of national policies of 'social good' (wages, employment, social services, protection of the environment), because such an approach raises costs. Since the national framework is, to date, the only one where enforcement is effective, this observation forecasts doom.

At the abstract, philosophical, level we are presented with a critique of meta-economics by John Broome. He suggests that of the three pillars of liberal economics - priorities dictated by limited resources; values determined by individual preference; the market as the tool for satisfying these preferences - only the first is of truly universal validity. The other two, if their domain of application is extended uncritically, may lead to quite significant deviations from the just goals preached by liberal economists.

Herman Daly argues from within the discipline of economics itself that growth may become gravely uneconomic. To the non-expert this may sound like a contradiction in terms. He argues that it is invalid to say that since the relative importance of the extractive part of the economy is in decline, concern for the environment is reduced. Nor is it tenable to say that whenever the interaction of production with nature becomes problematic, science or technology will be able to come to the rescue. In fact, the evidence seems to point the other way, bringing these theoretical conclusions into line with the paper by Edward Goldsmith and suggesting that growth may not be sustainable and is definitely not humane.

In the last paper, Nicos Mouzelis proposes a hermeneutic of modern society and its impasse by characterising the main features of 'modernity'. He draws attention to a series of concepts which can be interpolated between social science and mystery (in the sense used by the theologians in our symposium). In this, he emphasises the opposition of the Promethean-scientific-productivist approach (fragmenting and controlling) and the apophatic-religious-integrative approach. As in the previous papers, the spectre of Social Darwinism is raised as an integral and dominant component of the hegemonious political economy. The suggested antidote to the dangers of this 'modernist' transformation of society and neo-liberalism is democratisation within a more humane type of economic system. The author perceives the roots of such an economic system to lie within European capitalism.

It is appropriate that the chapter - with its otherwise gloomy consensus - ends on this optimistic note.

Trade and the Environment

Edward Goldsmith

Everywhere our forests are overlogged, our agricultural lands overcropped, our grasslands overgrazed, our wetlands overdrained, our groundwaters overtapped, our seas overfished, and just about the whole terrestrial and marine environment overpolluted with chemical and radioactive poisons. Worse still, our atmospheric environment is becoming ever less capable of absorbing either the ozone depleting gases or the greenhouse gases generated by our economic activities without creating new climatic conditions to which man cannot indefinitely adapt.

In such conditions, there can only be one way of maintaining the habitability of our planet and that is to methodically reduce the impact of our economic activities. Unfortunately, it is the overriding goal of just about every government in the world to maximise world trade and create a global economy. Economic development, by its very nature, must necessarily further increase the impact of our economic activities on our environment. This could not be better illustrated than by the terrible environmental destruction that has occurred in Taiwan which in the last decades has achieved the most stunning rates of economic growth.[1]

Forests have been cleared to accommodate industrial and residential developments and fast-growing conifers; the virgin broadleaf forests that once covered the entire eastern coast have now been almost completely destroyed and the vast network of roads built to open up the forests to logging, agriculture and development have caused serious soil erosion. Efforts to maximise agricultural production in export-oriented plantations have led to the tripling of fertiliser use between 1952 and 1980, and this has led to soil acidification, zinc losses and declines in soil fertility, with water pollution and fertiliser runoff contaminating ground water - the main source of drinking water for many Taiwanese. Pesticides are subject to no effective government control and are now a major source of contamination of Taiwan's surface waters and ground waters.

A substantial number of Taiwan's 90,000 factories have been located in the countryside, on rice fields along waterways and near private residences. To maximise competitiveness, their owners have disregarded what waste-disposal regulations there are and much of the waste is simply dumped into the nearest waterway. The government itself says that 20% of farmland is now polluted by industrial wastewater. In addition, agricultural and industrial poisons and human waste have severely polluted the lower reaches of just about every one of Taiwan's major rivers. In Hou Jin, a small town near the city of Kaohsiung, 40 years of pollution by the Taiwan Petroleum Company has not only made the water unfit to drink but actually combustible. Air pollution has also increased massively. Sulphur dioxide and nitrous oxide pollution in Taiwan are now intolerable, with levels regularly reaching double those judged harmful in the United States.

It could be argued that once Taiwan has achieved a certain level of Gross National Product, it will be able to install the technological equipment required for mitigating the destructiveness of the development process. Yet why should it, when much of the legislation that has been forced on recalcitrant governments by environmental groups in rich industrial countries is now being systematically repealed.

Production for export

A considerable proportion of world production of the most basic commodities is produced for export: 33% in the case of plywood, 84% of coffee, 38% of fish, 47% of bauxite and alumina, 40% of iron ore, 46% for crude oil.[2]

In Malaysia, more than half the trees that are felled for timber are exported. This brings in one and a half billion dollars a year in foreign exchange, but at a terrible environmental cost. Peninsula Malaysia was 70% to 80% forested 50 years ago. Today it has been largely deforested. The result has been escalating soil erosion, the fall of the water table in many areas, an increase in droughts and floods and the destruction of the culture and life-style of the local tribal people. As country after country is logged out, the loggers simply move elsewhere.

In Somalia, the export of sheep, goats and cattle has grown at least ten-fold and that of camel has increased twenty-fold since 1955 and by 1989 the export of livestock accounted for 42% of the country's foreign exchange earnings. The breakdown of the traditional nomadic system of rearing livestock has lead to overgrazing, soil erosion, and the degradation of range lands, limiting the ability of the land to provide sustenance for the Somali people.[3]

In mid-west America, the intensive cultivation of maize and soya beans, largely for export, is leading to such terrible soil erosion that what was once the most fertile agricultural area in the world will, at current trends, be almost entirely deprived of its topsoil within the next 50 years.[4] Tobacco accounts for 1.5% of total agricultural export and in the case of Malawi, represents 55% of that country's foreign exchange earnings. Damage is caused by the depletion of soil nutrients but the heaviest environmental cost of tobacco production lies in the sheer volume of wood needed to fuel tobacco curing barns. Every year the trees from an estimated 12,000 square kilometres are cut down, with 55 cubic metres of cut wood being burnt for every tonne of tobacco cured. The production of coffee also causes serious environmental degradation, as do groundnut plantations in French West Africa.

The fishing industry, itself dependent on exports, is having a similar effect upon the seas. Today, nine of the world's 17 major fishing grounds are in decline and four are already 'fished out' commercially.[5] In 1992, the great cod fisheries of the Grand Banks off Newfoundland in Canada were closed indefinitely and in Europe, mackerel stocks in the North Sea have decreased by 50 times since the 1960s. As fish stocks are depleted in the North, it is in the South that the fleets are now congregating, but the volume of fish exported from developing nations has already increased by nearly four times in the last 20 years and southern fisheries are already under stress.[6] The predictable result is the depletion of Third World fisheries, with drastic consequences for local fishing communities.

So far we have considered some of the local effects of extractive export industries. The produce of such industries, as well as that of mining minerals, oil, coal, natural gas and mass-produced manufactured goods must be transported to the countries that import them.

In 1991, four billion tons of freight were exported by ship world-wide, requiring 8.1 exajoules of energy, which is as much as was used by the entire economies of Brazil and Turkey combined. That year, 70 billion tons of freight were sent by plane, which used 0.6 exajoules, equal to the total annual energy use of the Philippines.[7]

A European task force has calculated that the creation of the single market in Europe in 1993 will increase cross-border traffic with a consequent increase in air pollution and noise of 30% to 50%. With the increase in trade between North America and Mexico, cross-border trucking has doubled in the last five years and this is before trade barriers were reduced between the two countries. The United States government predicted that after the signature of the North American Free Trade Agreement, cross-border trucking would increase by nearly seven times. The ratification of the GATT Uruguay Round Agreement can only further increase the world-wide transport of goods, and to accommodate it a vast number of new highways, airports, harbours and warehouses must be built, also causing serious environmental destruction. Extra combustion of fossil fuels for transport purposes will contribute to additional pollution and the risk of accidental spills of oil and dangerous chemicals during transportation will increase.

Indeed, if the environmental costs of increased transport were properly taken into account and 'internalised', it is likely that much of world trade would be seen to be totally uneconomic.

The environmental effects of increased competition

A recent European Commission report has seriously questioned the effectiveness of current environmental regulations noting that there has already been a 13% increase in the generation of municipal wastes between 1986 and 1991, a 35% increase in the EC's water withdrawal rate between 1970 and 1985, and a 63% increase in fertiliser use between

1986 and 1991. It predicts that if current growth rates continue, carbon dioxide emissions must increase by 20% by the year 2010, making nonsense of the European Union countries' commitment to stabilise them by the year 2000.

However, in the free-for-all of the global economy no country can strengthen environmental regulations that increase corporate costs without putting itself at a 'comparative disadvantage' with its competitors. For example, the European Union and Japan both proposed adopting an international tax on fossil fuels as a first step in a campaign to reduce carbon dioxide emissions. The United States refused, saying that imposing such a tax on Americans would be 'electorally impossible'. Not wanting to impose costs on themselves alone and thus render themselves less competitive, the European Union and Japan dropped the idea. Fossil fuel use and carbon dioxide emissions thereby remain almost entirely out of control.

Competition and environmental disaster

In order to increase competitivity, corporations increasingly undertake cost-cutting measures. The Exxon Valdez disaster would probably not have occurred if Exxon had not eliminated 80,000 jobs, reducing the crews of its supertankers by a third.[8] Normally the supertanker would have navigated in a safe but slow shipping lane; in order to cut costs, it was moved to a much faster, more dangerous lane which meant navigating through ice flows from the Columbia glacier.

Deregulation

Corporations are limited in their efforts to cut costs by regulations that protect the interests of labour, the unemployed, the poor, the old and the sick, and, of course, the environment. To the businessman, these regulations are so much bureaucratic red tape which increases costs and reduces competitivity. As a result, pressure has mounted everywhere to get rid of these regulations.

The likely effects of deregulation at a world level can be gauged from the experience of 'free trade zones' or 'export processing zones', of which there are now some 200 in the Third World, usually situated near key communication centres.[9] Foreign industries are enticed to establish themselves in these zones by the simple expedient of eliminating any effective regulations to protect the interests of labour or the environment. Wherever free trade zones have been established, there has been environmental devastation on a horrific scale.

Cross deregulation

More effective than deregulation carried out by national governments within their own country is that which is imposed on them by their trading partners under the GATT Uruguay Round Agreement. In the European Union's Report on US Barriers to Trade and Investment, April 1994, it is suggested that the commissioners should seek to overturn a large number of Californian and US Federal environmental laws which it is felt could successfully be classified as GATT-illegal trade barriers. These include California's Safe Drinking Water and Toxic Enforcement act (proposition 65) which requires the sticking of warning labels on products containing known carcinogenic substances, and the Nuclear Non-Proliferation act, a number of laws designed to protect fish stocks by limiting the use of large scale drift nets and other devices that lead to over-exploitation of fish stocks. It has been estimated that 80% of America's environmental legislation could be challenged thus. The United States and other countries can conveniently challenge European Union environmental laws in the same way.

The environmental effects of Structural Adjustment Programmes

The experience of those Third World countries that in the last ten years have been subject to International Monetary Fund and World Bank Structural Adjustment Programmes provides another eloquent illustration of the environmental consequences of increased competitivity and deregulation among export-oriented industries.

Costa Rica was subjected to no fewer than nine IMF and World Bank structural adjustment programmes between 1980 and 1989. Greatly increased exports were made possible by the massive expansion of the banana industry and of cattle ranching. The latter was heavily subsidised, a third of state agricultural credit going to the cattle ranchers. Expansion took place at the cost of the country's forest cover which dropped from 50% in 1970 to 37% in 1987 and has dropped still further ever since. To increase banana production, huge amounts of chemical fertilisers and pesticides have been used, which are then washed into the rivers and end up in the sea. In some areas 90% of coral reefs have been annihilated as a result.

Structural adjustment programmes have led to the same sort of environmental destruction in Ghana and in the Philippines, one of the most structurally adjusted countries in the world. In signing the GATT Uruguay Round Agreement we are in effect committed to subjecting the entire world to one vast structural adjustment programme.

Harmonising standards

Free trade has been institutionalised by a series of free trade agreements designed and promoted by associations of businessmen for whom environmental regulations are no more than costs that must be reduced to the minimum. Where possible the environmental issue has been avoided altogether. The word 'environment' appears nowhere in the mandate of the GATT nor is it mentioned in the constitution of the World Trade Organisation save in a cursory manner in the preamble.

Public pressure has forced the bureaucrats to take some notice of environmental issues and there is even some talk of 'greening the GATT'. Whatever the rhetoric, when it comes to adopting environmental standards that will increase costs to industry, they are invariably rejected. In 1971 the GATT secretariat stated that it was inadmissible to raise tariffs to take into account pollution abatement costs. In 1972 it refused to accept the 'polluter pays' principle.

Of course measures should be taken to ban or at least limit activities which are particularly destructive and channel economic development into those areas that are less so. But this is no longer possible. The global economy is controlled by increasingly stateless, unaccountable and ungovernable transnational corporations, which have set up, via the World Trade Organisation, a new international legal system that is designed to make it virtually impossible to adopt environmental controls that could increase their costs and thereby reduce their competitivity.

Creating consumers

Making a case for exporting the socially and environmentally devastating and utterly non-sustainable western lifestyle to the five billion or so people who have not yet entirely adopted it is impossible. It is mainly the appetite for this lifestyle that can be exported. The lifestyle itself only an insignificant minority will ever enjoy, and even then for but a brief period of time. It has been calculated that to bring all Third World countries to the consumption level of the USA by the year 2060 would require 4% economic growth a year. The annual world output, however, and the annual impact of our economic activities on the environment, would be 16 times what it is today, rendering the biosphere even less capable of sustaining the impact.

There is no evidence that trade or economic development are of any great value to humanity. World trade has increased by 11 times since 1950 and economic growth by five times, yet during this period there has been an unprecedented increase in poverty, unemployment, social disintegration and environmental destruction. The environment on the other hand is our greatest wealth. To kill it is an act of unparalleled criminality. It is also an act of unparalleled short-sightedness. There can be no international trade and no economic development on a dead planet.

Notes

1. Bello, W., Rosenfield, S. (1990) *Dragons in Distress*, Institute for Food and Development, San Francisco.
2. French, H. (1993) Costly Tradeoffs, Reconciling Trade and the Environment, Worldwatch Institute, Washington DC.
3. French, H., Worldwatch Institute.
4. Krohe, H. (1984) Illinois - The US Breadbasket. Where has all the soil gone? *The Ecologist*, Vol. 14 No5/6.

5. The editors of *The Ecologist*, Vol.25, Nos.2-3, 1995, p.42.

6. French, H., op. cit.

7. French, H., op. cit.

8. Hawken, (1993) *The Ecology of Commerce*, San Francisco, 1993.

9. Goldsmith, A. (1996) *Free Trade Zones*, Sierra Club Books, San Francisco.

Economics and the Environment

John Broome

I am a lapsed economist. I used to work as an economist, but I now work as a philosopher. This is partly because I am not happy with a lot of economics as it is practised, and therefore I cannot present the arguments of economists without qualification.

In this quick tour of capitalist ideology and its errors I shall outline the case for free markets and shall then point out the places where I think it goes wrong.

There are three premises to the standard arguments from economics:

1. Resources are limited, and not sufficient for everything that would be desirable. It is necessary therefore to determine priorities and make compromises. We have to make judgements about the comparative values of alternative uses of resources.
2. The comparative values of these alternative uses are properly determined by people's own desires and preferences. That is to say: what is worth pursuing is what people actually care about. This is liberalism. Value comes from the attitudes of individual people, not from any authority.
3. The best way of satisfying desires and preferences is by means of a free market. There is a lot of theory behind this, but the idea is simple: if two people make an exchange of goods, each of them may benefit, and neither can lose. Neither can lose because neither is compelled to make the exchange if he or she does not want to. To take a controversial example, if I can spare a kidney and need money to send my children to school, and if you need a kidney and can spare some money, we can both benefit by my exchanging my kidney for your money.

The conclusion is that markets should be as free and extensive as possible.

Now may I offer some comments on the premises, in reverse order.

3 (a): *Externalities*. Markets do not always capture the true value of things, because some of the people who gain or lose by an exchange may not participate in the agreement. When I trade oil with you, neither of us takes account of the people who have to live with the mess we make along the way. The problem of externalities is well-recognised by environmental economists, and the usual response is to make sure the external costs and benefits are made to impinge on the market. The principle of 'polluter pays' is one way of doing this. Another is to impose taxes on goods that produce externalities, equal to the harm done by the externalities. Ideally, if externalities are properly internalised in one of these ways, markets should then serve their

purpose properly. One aim of environmental economics is to bring about this ideal result, but there are many difficulties in the way. An important difficulty is that we often do not know the size of the external costs and benefits, so we do not therefore know what taxes to impose. To a large extent the hope of an ideally functioning market is a fantasy.

(b): *Monopolies*. If there are monopolies that exercise power in a market, the market does not function as it is supposed to. When huge multinational companies and cartels are trading with poor countries and individuals, they clearly disrupt the proper functioning of the market.

(c): *Unequal representation*. Markets do not truly satisfy everyone's desires and preferences, but only those that are backed by purchasing power. The poor get their preferences poorly satisfied. Moreover, the market itself determines people's purchasing power: it gives little to people who have few resources. One major group of people without purchasing power are those who have not yet been born. Future generations are not represented in the market at all, yet they are the main victims of pollution.

2. The second premise is simply false. It is false that the real value of things is determined by people's desires and preferences. Sometimes people desire what is truly valuable, but sometimes not. The evils of consumerism represent one example: people may desire consumer goods that are not actually worth having. The truth is that our desires and preferences ought to be determined by the real value of things, not that the real value of things is determined by our preferences. The opposite, false, view is very deeply embedded in the thinking of nearly all economists.

1. The first premise is true, and an important insight from economics. We cannot do everything; we have to compromise. Nature is valuable, but so are people. Sadly, people's needs cannot be fully harmonised with those of nature. We shall often have to weigh the interests of people against the interests of nature, and it is right to do this weighing. People need food, which requires fertiliser; they need medicines, and they need many other things, including some fun. The extinction of a species is bad in itself, but it may be right to permit it if the costs to people of saving it are too large. We must recognise the need for compromise, and the methods of economics are helpful for making compromises. Compromise is, in a way, the whole subject matter of economics, and since compromise is never popular, economics has a gloomy image. We do need reviled methods such as cost-benefit analysis but they must be used properly. Often they are not. They should not rely only on market prices, and values derived from people's preferences. Principally, they must take into account the true value of things.

Uneconomic Growth: Conflicting Paradigms

Herman E. Daly

Uneconomic growth in theory

Growth in Gross National Product is so favoured by economists that they call it 'economic' growth, thus ruling out by terminological baptism the very possibility of 'uneconomic' growth in GNP. There is, however, no *a priori* reason why at the margin the environmental and social costs of growth in GNP could not be greater than the production benefits. In fact, economic theory would lead us to expect that at some point. The law of diminishing marginal utility of income tells us that we satisfy our most pressing wants first, and that each additional unit of income (production) is dedicated to the satisfaction of a less pressing want. So the marginal benefit of growth declines.

Similarly, the law of increasing marginal costs tells us that we first make use of the most productive and accessible factors of production - the most fertile land, the most concentrated and available mineral deposits, the best workers - and only use the less productive factors as growth makes it necessary. Consequently, marginal costs of growth increase.

When rising marginal costs equal falling marginal benefits then we are at the optimal level of GNP, and further growth would be uneconomic. It would increase costs more than it increased benefits. Why is this simple application of the basic logic of microeconomics treated as inconceivable in the domain of macroeconomics?

Uneconomic growth in fact

One might accept the theoretical possibility of uneconomic growth, but argue that it is irrelevant for practical purposes since, it could be alleged, we are nowhere near the optimal scale. The benefits of growth might still be enormous and the costs still trivial at the margin. Economists all agree that GNP was not designed to be a measure of welfare, but only of activity. Nevertheless they assume that welfare is positively correlated with activity so that increasing GNP will increase welfare, even if not on a one-for-one basis. This is equivalent to believing that the marginal benefit of GNP growth is greater than the marginal cost.

Uneconomic growth in two paradigms

Within the standard neoclassical paradigm uneconomic growth is an anomalous category. You will not find it mentioned in any macroeconomics textbook. But within the paradigm of ecological economics it is an obvious possibility. Let us consider why in each case.

The preanalytic vision of standard neoclassical economics is that the economy is the total system, and that nature, to the extent that it is considered at all, is a sector of the economy, for example, the extractive sector (mines, wells, forests, fisheries, agriculture). Nature is not seen, in the vision of ecological economics, as an envelope containing, provisioning, and sustaining the entire economy, but as one sector of the economy similar to other sectors. If the products or services of the extractive sector should become scarce, it is presumed the economy will 'grow around' that particular scarcity by substituting the products of other sectors. If the substitution is difficult, new technologies will be invented to make it easy.

In this view, the unimportance of nature finds empirical support in the declining share of the extractive sector in total GNP. Beyond the initial provision of indestructible building blocks, nature is simply not important to the economy in the view of neoclassical economics. Ecological economics considers the percentage of GNP represented by resources to be a misleading indication of their importance. One might as well claim that a building's foundation is unimportant because it represents only five percent of the height of the skyscraper erected above it. GNP is the sum of value added. Resources are that to which value is added - the foundation or base upon which the skyscraper of value added is resting. A foundation's importance does not diminish with the growth of the structure that it supports! If GNP growth resulted only from increments in value added to a non-growing resource throughput, then it would remain economic growth. That is not what happens.

What happens, according to ecological economics, is that the economy grows mainly by transforming its environment (natural capital) into itself (man-made capital). This process of transformation takes place within a total environment that is finite, non-growing, and materially closed. A throughput of solar energy powers biogeochemical cycles, but that energy throughput is also finite and non-growing. As the economic subsystem grows it becomes larger relative to the total system, and therefore must conform itself more to the limits of the total system - finitude, non-growth, and entropy. Subsystem growth is ultimately limited by the size of the total system, even under neoclassical assumptions of easy substitution of man-made for natural capital. But if man-made and natural capital are complements rather than substitutes, as ecological economics claims, then expansion of the economic subsystem would be much more stringently limited by that complementarity. There would be no point in transforming natural capital into man-made capital beyond the capacity of remaining natural capital to complement and sustain it. What good are more fishing boats when the fish population has disappeared? The fish catch used to be limited by the number of fishing boats (man-made capital) but is now limited by the remaining populations of fish in the sea (natural capital).

When factors are complements, the one in short supply is limiting. If factors are substitutes then there cannot be a limiting factor. Economic logic says that we should economise on and invest in the limiting factor. Economic logic stays the same, but as we have moved from an 'empty' world to a 'full' world, the role of limiting factor has gradually shifted from man-made to natural capital, from fishing boats to remaining fish in the sea; from sawmills to remaining forests; from irrigation systems to aquifers or rivers; from oil well drilling rigs to pools of petroleum in the ground; from engines that burn fossil fuel to the atmosphere's capacity to absorb CO^2.

The smaller the optimal scale of the economy, the greater is: (a) the degree of complementarity between natural and man-made capital; (b) our desire for direct experience of nature; (c) our estimate of both the intrinsic and instrumental value of other species. The smaller the optimal scale of the economy, the sooner its physical growth becomes uneconomic.

From permitting growth, to mandating growth, to limiting growth

The neoclassical paradigm permits growth forever, but does not mandate it. Historically the growth mandate came from the answer given to the problems raised by Malthus, Marx and Keynes. Overpopulation, unjust distribution, and involuntary unemployment would all be solved by growth. Overpopulation would be cured by the demographic transition initiated by growth. Unjust distribution of wealth between classes would be rendered tolerable by growth, the rising tide that lifts all boats. Unemployment would yield to increasing aggregate demand which merely required that investment be stimulated, which of course implies growth.

Continuing this time-honoured tradition the World Bank's 1992 World Development Report argued that more growth was also the solution to the environmental problem. The assumption in all cases was that growth was economic, that it was making us richer rather than poorer. Now growth is becoming uneconomic. Uneconomic growth will not sustain the demographic transition and cure overpopulation. Neither will it help redress unjust distribution, or cure unemployment. Nor will it provide extra wealth to be devoted to environmental repair and clean-up.

We now need more radical and direct solutions to the problems of Malthus, Marx and Keynes: population control to deal with overpopulation; redistribution to deal with excessive inequality; and ecological tax reform to raise resource productivity and employment. These must be national policies. It is utopian (or dystopian) to think of them being carried out by a world authority. Many nations have made progress in controlling their population growth, in limiting domestic income inequality, in reducing unemployment. They have also improved resource productivity by internalising environmental and social costs into prices. These significant national gains are now being undercut by the ideology of globalisation. Global economic integration by free trade and free capital mobility effectively erases the policy significance of national boundaries, turning the federated community of nations into a cosmopolitan noncommunity of globalised individuals. Some of these 'individuals' are giant transnational corporations, treated as individuals by legal fiction.

Under globalisation, each country seeks to expand beyond the limits of its own ecosystem and market by growing into the ecological and economic space of all other countries, as well as into the remaining global commons. Globalisation operates by standards-lowering competition to bid down wages, to externalise environmental costs, and reduce social overhead expenses for public goods. It is far worse than an unrealistic global dream - it actively undercuts the ability of nations to continue dealing with their own problems of unjust distribution, unemployment, external costs, and overpopulation. It is hard to imagine any country continuing to limit its birth rate or internalise its environmental and social costs when the results of overpopulation and cost externalisation in other countries freely spill over into it.

Globalisation is the latest elixir concocted by the growth-forever alchemists. Export-led growth is the new philosopher's stone; it turns lead into gold by the alchemy of free trade. With the revival of alchemy comes a return to the logic of mercantilism: wealth is gold, and the way for countries without mines to get gold is to export more goods than they import and receive payment for the difference in gold. The way to export more than you import is to reduce wages. The way to keep wages low is to have an oversupply of labour, attained by easy immigration or high birth rates among the working class. Globalisation requires, therefore, that for a nation to be rich, the majority of its citizens must be poor, increase in number, and live in a deteriorating environment.

Truly, John Ruskin foresaw the era of uneconomic growth, a time when: 'That which seems to be wealth may in verity be only the gilded index of far-reaching ruin' (*Unto this Last*, 1862).

Religion, Science and the Environment: a Synthetic View

Nicos Mouzelis

I have been helped in making sense of the environmental and religious issues we are dealing with by using a concept which, although contested and polysemic, plays a crucial role in the social sciences today: the concept of Modernity.

One way of defining Modernity is to view it as the type of social organisation which spread widely in western Europe after the Industrial and the French Revolutions. From a sociological point of view, it has two unique characteristics.

First, Modernity is characterised by unprecedented levels of social mobilisation as administrative and communicative technologies penetrate the periphery of society, undermining traditional localisms and bringing the majority of the population into the broader economic, political, social and cultural arenas of the nation-state. One can grasp the uniqueness of the modern predicament if one contrasts the very few links between 'subjects' and the pre-industrial state with the myriad daily links that citizens have with the modern state via schools, the post-office, the national systems of communications and the like.

The other unique characteristic of Modernity is an equally unprecedented degree of institutional differentiation as economy, polity, society, culture and religion tend to acquire distinct dynamics, distinct values and distinct rationalities. One does see a similar type of differentiation in complex pre-industrial societies, but in these societies social differentiation is limited to the top. The bulk of society is still organised in a non differentiated, segmental manner.

The four dimensions of Modernity

One can identify four major dimensions of Modernity, particularly in relation to the environment.

The cultural dimension is characterised by Prometheanism, the term used by the Bishop of London,[1] a state of affairs where via massive secularisation we have arrived at what Max Weber has called the disenchantment of the world.[2] The cosmos loses its sacred character, human beings are cut off from the divine – while at the same time they compensate for their lost security by viewing themselves as masters of the environment and the cosmos.

This Promethean idea of mastery was, of course, enhanced by the seventeenth century scientific revolution which, for the first time, systematically linked scientific exploration with the so-called practical arts and crafts. We thus arrive at the present radical desacralisation of the environment and its reduction into a huge machine for the satisfaction of humankind's unbounded wants and ambitions.

In the economic sphere, Modernity is characterised by the complex Industrialism / Productivism. The development of applied science led to the large-scale spread of the factory system and the extraordinary development of the means of production in all economic spheres. Whether these new, formidable forms of production were controlled by capitalism or state planning (as in the ex-Soviet Union), the mode of control was, and is, based on the idea of the unlimited, unbounded production of goods as the only means for achieving human happiness and security. We have thus arrived at the present situation where the logic and values of productivity and economic competition reign supreme.

In the political sphere, the most striking feature of Modernity is the shifting of people's loyalties and orientations from their local traditional communities to the national centre. This has led to the emergence of nationalistic ideologies and the extraordinary development of the means of violence, and their control, by national élites operating in a system of antagonistic nation-states; nation-states whose rivalries led to the unprecedented human and environmental catastrophe of the two World Wars.

Finally, in the social sphere, Modernity, particularly in its present neo-liberal Thatcherite phase, is characterised by a new type of Social Darwinism (at least on the level of practice if not on that of rhetoric). In pre-modern times, privileged members of the community were supposed to have not only rights, but also obligations to their less fortunate fellows. This aristocratic type of solidarity was replaced by the idea that since the market works for the good of everybody, those who do not succeed economically have only themselves to blame. Social Darwinism was mitigated during the first three post-war decades in western Europe. Trade unions and social democratic parties managed, if not to save the environment, at least to humanise capitalism by the development of the welfare state, by state/capital/labour collaboration, and by the adoption of Keynesian means for the overall regulation of the national economy. However, since the 1974 world economic crisis and the rapid globalisation which followed, Keynesian means do not seem to work any more in a situation marked by growth in unemployment, the crisis of the welfare state, the creation of the two-thirds/one-third society, and the growing polarisation between rich and poor countries globally.

This combination of Prometheanism, Productivism, Nationalism, and Social Darwinism can give us an overall explanation not only of large-scale environmental destruction, but also of the replacement of what Martin Buber termed the 'I-Thou' relationship[3] with an instrumental 'I-It' relationship. In other words, the Promethean desire to seek security and happiness via domination and utilitarian control characterises not only the Self-Nature relationship but also the Self-Other and even the Self-Self relationships.

The relevance of democratisation

In this symposium the holistic character of theology has been stressed, in contrast to the fragmented approaches of the natural sciences. In the social sciences as well - particularly in their positivistic version, which is still dominant today - social phenomena are studied by disciplines which communicate very little with each other. However, in order to study environmental problems successfully, this type of compartmentalisation has to be overcome. We need a more holistic, integrated, conceptual framework which will take seriously into account not only the technical or even spiritual/cultural, but also the economic, political and social dimensions of the environmental crisis. It is only in this way that environmental problems can be properly defined and the limits of present environmental strategies properly assessed.

Let me make this crucial point clear by providing some examples. Karl Popper pointed out decades ago that it is impossible to predict the future direction of scientific developments because, by definition, future scientific knowledge cannot be known in advance. However, it is also true that scientific and technological developments are not only shaped by the inner logic of the scientific enterprise, they are also shaped by the amount of resources that we assign to different types of applied research.

We can develop this point by making a distinction between Promethean technologies and - borrowing a theological term - apophatic technologies. Promethean technologies try to transcend or transgress the limits imposed by our inner and outer nature. Apophatic technologies try to restore natural balances and harmonies. They try to remove or peripheralise those factors which undermine the logos and harmony that we find in nature. Genetic engineering, particularly when oriented towards eugenic goals, is a good example of the Promethean type of technology, and anti-pollution technologies are good examples of the apophatic type. Whether technological development will take a more Promethean or a more apophatic direction has very much to do with politics, particularly with the democratic process. If the decision-making process in matters related to applied science is constructed in such a way that the interests of the majority are clearly articulated and represented, it is likely that we are going to move in an apophatic direction. If on the other hand corporate or other élite interests prevail, we are going to continue in the present Promethean mode. Therefore, the issue of radical democratisation is not peripheral, but rather constitutes the core of the environmental problem.

Two more points illustrate the crucial, constitutive role that moral and political economy should have in the environmental debate. First, we are aware of the importance of environmental education and the need for a massive change in values and consumption patterns if we are going to cope with the present crisis. However, we cannot talk seriously about a massive change in values or lifestyles without taking into account that we live in a world where economic capital can, to use P. Bourdieu's terms, automatically buy cultural or symbolic capital.[4] This means, in fact, that a few global media-owners have more influence on identity formation and value creation than all the environmentalists, religious leaders, teachers and artists put together.

Secondly, as Ulrich Beck has argued in his famous book *The Risk Society,*[5] the most serious danger to the environment comes from the fact that crucial decisions affecting it are taken in laboratories and business boards whose proceedings are outside the public domain, and therefore far removed from democratic control.

Thus the solution of the environmental problem is inextricably linked with what Beck has called ecological democracy, that is with the need to radically deepen the democratisation process both nationally and globally.

The neo-liberal global order

The need for a holistic, multi-dimensional approach does not apply only to the nation state, but also to the global system, whose neo-liberal mode of organisation can also be adequately described in terms of the four institutional features of Modernity discussed above.

The so-called New World Order is based on the Promethean and at the same time Productivist belief that through the spread of western technology, capital and culture, all countries will sooner or later industrialise in a linear manner - moving up the evolutionary ladder. It is also founded on an arrangement which clearly favours certain nationalist interests at the expense of others. Finally, this global order is legitimised by the idea that through market mechanisms, not only is the maximum of wealth produced, but also, via 'trickle down' processes, it will eventually spread to everybody. Given the fact that the famous 'trickle down' process does not occur, this ideology simply hides a Social Darwinian attitude, in which the implicit thesis is that the marginalisation of large sections of humanity is the price that we have to pay for 'progress'. Those who focus on the terrible conditions of poverty in Brazil and elsewhere go straight to the heart of the problem. The destruction of the environment is inextricably linked with the problem of the marginalisation of a large part of humanity by the neo-liberal global order. If we focus only on the environment and ignore the issue of marginalisation, we simply risk becoming irrelevant.

Looking at the performance of the neo-liberal global order, one has to admit that the market dominated world economy does produce more wealth than alternative forms of economic organisation. However, we also have to admit that this wealth is badly produced and badly distributed, for it destroys the environment while at the same time it excludes the majority of the world's population from the beneficial effects of development. From this point of view, the neo-liberal world order constitutes a system of organised irresponsibility or a system of collective greed. It is a system which is not only profoundly immoral, but also profoundly short-sighted and even stupid.

Edward Goldsmith has forcefully argued that the global system is so dominated by the multinationals that it leaves no room for any serious reforms in the environmental or the social spheres. Although I accept the thesis of multinational dominance, and admire his holistic, institutional approach to global issues, I disagree with him on two points. First, even within the existing limitations, I think serious work has been and can be further done towards a more sustainable and socially just development. Secondly, I think that the limits imposed by the global economy can be broadened within capitalism, for capitalism takes different forms, and some forms are better than others. The so-called Rhine model of capitalism, which stresses investment in human capital, is more humane than American capitalism, with its emphasis on down-sizing and maximising profits in the short term. Moreover the Rhine model, prevalent in western Europe, is more democratic than Asian capitalism, which has achieved its dynamism by destroying the environment and repressing working class organisations. At the moment the social democratic European type of capitalism is on the defensive, but it is not dead and can become much stronger if Europe achieves greater unity, and therefore greater capacity to put its stamp on the world economy.

Moreover, one should not forget that the neo-liberal character of the global system is precarious for another reason. The rapid industrialisation of China will very soon destroy the Promethean myth of unbounded productivism. Sooner or later, political and economic leaders will have to face the fact that 'business as usual' for large scale industrialisation will be physically impossible. Some form of stricter regulation of the world economy will have to come about. The challenge, of course, is to make sure that the regulation to come will have a more democratic and humane character.

We should strive for a shift from the present barbaric, neo-liberal system to a more civilised global order. In this new order economic productivity will not dominate but will coexist in a balanced manner with genuine democracy in the political sphere, solidarity in the social sphere, and respect for the sanctity of the human person and nature in the cultural and religious spheres.

Notes

1. See Chapter 3.

2. Weber, M. (1978) *Economy and Society*, eds. Roth G., Wittich, C. University of California Press.

3. Buber, M. (1937) *I and Thou*, T. and T. Clark, Edinburgh,.

4 Bourdieu, P. (1988) *Homo Academicus*, Polity Press, Cambridge.

5. Beck, U. (1992) *The Risk Society*, Polity Press, Cambridge.

Recommendations for Helping to Protect the Black Sea

Introduction

The symposium made an important contribution towards a more general understanding of the environmental crisis affecting the Black Sea and the coastal areas beyond its shores. However, many of the speakers recognised that mere understanding did not go far enough. Bringing the knowledge of science and the wisdom of religion together as a joint force should enable a new approach to tackling many of the environmental problems at source - in the failure of humans to be good stewards of nature. Many proposals were made during the symposium, a number of which were strikingly similar. We have condensed the conclusions of this work and hope that it may be followed up by all who participated in this event or who have taken the time to study this book.

One point that is important to clarify from the outset: the symposium did not pretend to replicate the Black Sea Strategic Action Plan (see Box 1) which was signed by the six Ministers of the Environment of the Black Sea countries on 31 October 1996. Indeed it gave its full support to this excellent strategy and participants expressed their strong desire that the Black Sea Strategic Action Plan should be implemented without further delay. The actions presented here complement the Action Plan and offer new insights into practical activities for achieving its objectives.

The actions proposed by the symposium fall into four general groups: actions to enhance governance; development and use of the public media; enhancing environmental education; and actions to encourage participation of civil society in environmental issues.

Enhancing governance

Issues

A number of strong commitments have been made by governments in the Black Sea region for programmes which would help to save the Black Sea from further degradation and to restore it to a diverse and healthy state. Unfortunately, these programmes are not being fully implemented. The problem would seem to be due to the weak position of Ministries of the Environment in central governments but the issue is in reality more complex. In many cases, the role of regional and local governments is poorly defined and foreign aid, channelled through central governments, has little impact at the local level. Much more work is needed at the local community level if democratic structures are to be enhanced. The Black Sea may be a very small part of a plethora of problems being tackled by central governments but is a much more significant issue in coastal communities whose livelihoods depend upon the health of the sea.

The Black Sea Environmental Programme (BSEP) is currently coordinating international efforts to save the Black Sea. The BSEP has been successful in building capacity in the Black Sea on a wide range of issues and in helping governments to develop new policies and to lever the necessary financial investments to put some of them into effect. As participants in the symposium appreciated, some of the countries in the region are facing difficult financial circumstances in their transition to western-style democracies and market-based economies. There is even less to spare for international environmental work such as the BSEP.

Actions proposed by the symposium

The symposium initiated a call to international donors to continue funding for the BSEP. The initial response from the European Union, the Global Environmental Facility programme and the World Bank has been very positive. This does not imply however, that the problem has been resolved as the rules of the Global Environmental Facility and the World Bank do not enable further direct support. Donors were particularly invited to enter partnership agreements with the governments in the region.

Coordinated donor support is urged to local communities. This is work where all sectors can participate: business leaders to invest and improve the environmental consciousness of local businesses and the advantages of 'going green'; local authorities to seek donor support to exchange experiences of improved environmental policies; NGOs in the West to exchange experiences and know-how with those in the Black Sea region; and religious leaders to work with local communities on environmental protection. This last issue is of particular importance. The symposium highlighted the protection of the environment as a spiritual issue and not just a technical one. Indeed, it demonstrated that part of our failure to protect the environment is a failure in our own system of values and a lack of appreciation of the beauty of God's creation.

Coordination of scientific programmes in the Black Sea region is also tenuous and the existing excellent programmes (initiated by NATO and by the Intergovernmental Oceanographic Commission of UNESCO) need to be institutionalised. Furthermore, the symposium network is a powerful and unique tool, bringing together scientists and religious leaders from all over the world and extending deeply into the local communities in the region. Maintenance of an active network is essential to any coordination programme to save the Black Sea and support from symposium participants is therefore essential.

The symposium also identified the urgent need to bring together a wide sector of spiritual leaders, scientists and members of political and civil society from the Danube basin, in order to examine their role in understanding and protecting the Black Sea environment through protection of this major European river. It therefore proposed to hold a new symposium, Symposium III, 'A River of Life', in 1999, on a vessel travelling down the Danube from Germany to the Black Sea, based on the experiences of the past symposia.

There is a need to maintain international attention on the plight of the Black Sea and to enhance understanding that this is not an issue where a single simple solution can be applied, but is related to the complexity of nature itself. Western participants in the symposium were invited to investigate their own government's policy towards protection of the environment in the Black Sea region and to bring the issues identified by the symposium to their attention. Participants from the region were invited to examine how their governments were implementing the Black Sea Action Plan, and to bring the Plan to the attention of their own local authorities, urging them to participate fully.

Development and use of the public media

Issues

In the past five years, the public media has developed in a spectacular manner throughout the Black Sea region. People are spending increasing amounts of their time watching television and it is having an enormous influence on their awareness and attitudes. Environmental reporting, however, is still in its infancy. Few journalists are able to dedicate their work to environmental issues and few scientists take the time to explain their work to journalists. There are very few inspiring, quality, environmental documentary films produced inside the Black Sea region - viewers are more likely to see wildlife programmes from the West than programmes about their unique local ecosystems. Interactive programmes such as chat shows on environmental matters have begun in some countries but they often lack materials or know-how to captivate the audience. Investigative journalism sometimes confronts barriers of press freedom or civil rights.

Actions proposed by the symposium

There has been a large amount of material produced by international and local journalists regarding the Black Sea. Much remains unedited. This material could be compiled or re-edited in a manner which allows it to be distributed at low cost to the television channels in the region. This work could be highly cost-effective but it requires the development of a proposal and budget for submission to donors.

Small training courses need to be set up for environmental journalists. Members of the profession rightly do not expect to be told what to write or what to point their cameras at but nevertheless can benefit from interaction in small groups with successful environmental journalists from outside the region. They can also benefit from opportunities to spend 'quality time' with scientists in order to deepen their understanding of the technical background to environmental issues. These sessions may be organised on an individual basis or structured into workshops to which members of the press can be invited. The workshops need to be designed by journalists working with scientists.

Religious leaders and environmentally aware members of civil society could dedicate time for communicating their own wisdom on environmental issues. Speaking to the press is an important way of reaching a wider audience.

Writing to the local newspaper is an effective means for individuals to express their concern. Few people take the time but many people read the letters. Local TV or radio stations or the news desk of a local paper could be requested to provide greater coverage of the Black Sea crisis.

Enhancing environmental education

Issues

The crisis in the Black Sea's ecosystem is part of a planetary crisis in which we are rapidly becoming aware that 'business as usual' is unsustainable. Whilst there is wide agreement that many of the practices associated with our current lifestyles are leading to the destruction of our natural ecosystem, there is a general reluctance to change them. To cherish nature for its own sake, as well as for its provision of commodities and habitat, implies affording it a special place in our system of values. Abject utilitarianism in our society reflects this ethical crisis. We are failing to take care of our environment and to share resources equitably.

During the symposium, it was stated that, 'In the Black Sea region and its catchment, efforts are needed to educate the general public with clear, practical messages about conservation and sustainable development'. The younger generation should be one focus of the energy that we are able to bring to bear on this important task. The formation of sound environmental values and attitudes is not something which necessarily occurs in the family setting. In the West it is common for children to learn at school about changing practices and lifestyles to protect the environment and to challenge their parents at home. Western society is gradually awakening to the importance of environmental education in its widest sense, including both in the formal setting of the classroom and through the direct contact with nature that engenders deeper feelings. Both kinds of educational experience are needed in the countries around the Black Sea as they experience new pressures which encourage individualism and utilitarianism. At this time, the move towards environmental education in schools relies on the commitment and enthusiasm of a few inspired school directors who are trying to unite their efforts in a concerted manner, and their task is made more difficult by limited funding for education.

Religious leaders concerned with the creation of a more spiritual and caring society and who realise that developing a concern for the environment is a part of that process, face similar challenges. In both domains there need to be efforts to increase knowledge, understanding, experience and resources.

Actions proposed by the symposium

(a) Environmental education for and with the Clergy

At the initiative of His All Holiness The Ecumenical Patriarch Bartholomew I, environmental seminars have been held each summer for five years. On the basis of the symposia experience, the focus of the seminars over the next five years is being directed towards realising to its fullest the potential collaboration between religious and scientific leaders, and ensuring that the clergy have what they need to become actors and leaders in those communities whose concerns about the environment are growing. This will be launched through the establishment of a new form of summer seminar or institute. Participants will include clergy, teachers and students from the countries of the Black Sea region, scientists and environmental educators and additional representatives from educational, environmental and religious institutions. Participants will engage in environmental fieldwork, hands-on scientific and mathematical activities, spiritual exercises related to the sacredness of the earth and the development of plans for effectively continuing work back in their home areas. The five year effort will be grounded in the summer institutes but will develop to include communication networks, seminary curriculum development and the establishment of linked local centres for education, training and practice.

(b) Environmental education in schools of the Black Sea region

There are scattered pioneering efforts by individual teachers and school directors in the Black Sea region to make environmental education part of their own curricula. A number of teachers have independently expressed the need to form a network to support their initiatives with access to the international community of environmental educators. This was established shortly after the symposium. However, if it is to continue functioning and to increase its potential, outside support is needed. A number of plans have already been formulated, for example the production of teaching materials about the Black Sea. Many schools need specialised equipment and training. Current capabilities vary considerably - some of those involved already have sophisticated Internet-based teaching systems, others are unable to afford new textbooks or even a telephone connection. The establishment of a Black Sea Environmental Education Trust would assist the teachers in their work.

A series of summer environmental education camps are needed to promote direct contact of children with the natural environment and allow them to deepen their understanding of environmental issues through personal experience.

There is also a need for environmental education resource centres. These could be based upon existing infrastructure in the region and modelled on similar centres established in western Europe and North America. They would be self-sustaining in the medium term but will require initial donor support from outside the region in order to start functioning. State education authorities should be encouraged to support these centres and to gradually amend school curricula to include environmental education.

Encouraging participation of civil society in environmental issues

Issues

Individual and corporate action to protect the environment should be an intrinsic part of the development of a pluralistic society. In the Black Sea region, there are very few ways to stimulate the direct involvement of any sector in the hard work of environmental protection. Recognition of special services rendered to environmental protection has been an effective tool in many cultural settings throughout the world. Green awards for industry have been employed in the promotion of the industry itself. Recognition of individuals engaged in environmental work in sectors where it has a low profile (journalism, for example) raises the respectability of being a good environmentalist.

The involvement of the nascent private sector is particularly important in the case of the Black Sea. There is currently enormous development pressure for the Sea to be used as a major oil shipping route and such development rightly deserves to be regarded with caution as it may severely compromise other users of the Black Sea, including its natural flora and fauna. Business partnerships which follow clear principles for sustainable use may encourage practices which are in greater harmony with environmental protection. Environmentally friendly tourism is being discussed in terms of such partnerships but there is no progress in other areas. However, the presence of business leaders in the symposium was clear evidence of their interest in exploring a more active role in environmental protection.

Actions proposed by the symposium

(a) Black Sea Environmental Awards

The purpose of the awards, under the auspices of the Patriarch, is to recognise the accomplishments that contribute to the protection of the Black Sea. Examples include:

1. An award to business and industry leaders to recognise those whose practices have demonstrated a commitment to sustainable development and who have encouraged their colleagues to follow suit;

2. An award to architects, engineers and developers who demonstrate through the example of built projects a commitment to sustainable resource use and environmental conservation;

3. A monetary award to recognise excellence in environmental reporting through writing, broadcast, and/or film;

4. An award to recognise a 'Young Environmental Scientist(s) of the Year'.

In all cases, a transparent, impartial and open process would be developed to ensure the highest possible credibility for this scheme. Sponsors are needed to support the awards.

(b) Environmentally responsible business partnerships

There are some important recent precedents for the creation of an ethical code of practice to underpin good business practice. The gradual change in the attitude of many western manufacturers towards reducing waste production offers considerable hope for the future and will lead to positive benefits to all parts of the environment, including the sea and coastal areas. One of the most important schemes brought to the attention of the Religious and Scientific Committee is *The Natural Step*, an initiative begun in 1989 by Karl-Henrik Robert, the Swedish cancer researcher, and now widely accepted in western countries. He achieved a consensus between a group of prominent scientists and industrialists on four system conditions that determine whether or not an activity is sustainable: materials from the earth's crust must not systematically increase in nature (for example, heavy metals, fossil fuels); persistent substances produced by society must not systematically increase in nature (for example PCBs, CFCs, DDT); the physical basis for the earth's productive natural cycles and biological diversity must not be systematically deteriorated; and there must be a fair and efficient use of resources with respect to meeting human needs.

The four Natural Step principles are an important practical application of the precautionary approach. Their adoption by a group of businesses in the Black Sea region would represent an important covenant which would be difficult for the wider business community to ignore.

Conclusion

The symposium is part of a continuing dialogue on environmental protection in the Black Sea which has resulted in action by individuals or groups of individuals to cherish and protect this unique environment. Nobody is excluded; everyone has a direct role.

The actions presented here are not exhaustive. Each of them reflects the spirit of the symposium and are designed to demonstrate, in a very practical manner, how religion and science, business and politics can work in harmony and bring their unique qualities to bear on the problem of the Black Sea. Indeed, the inspiration of the great world religions reminds us that a small group of committed individuals, united in their beliefs and divine calling, can change the world we live in and restore hope for a better future.

Box 1. The Strategic Action Plan for the Rehabilitation and Protection of the Black Sea[1]: twenty major points

- adoption of a new institutional framework for the Black Sea, building on the achievements of BSEP and including the creation of a project implementation unit within the Secretariat of the Istanbul Commission;
- encouragement of a basin-wide approach to certain policy areas;
- implementation of profound fiscal reform - the implementation of the polluter pays principle for pollution source control by 1999 - through the adoption of permitting and licensing procedures addressing common Water Quality Criteria for specific types of water use;
- abatement of priority hot spots by the year 2006 - public progress reports in 2000 and 2005;
- development of specific plans for waste water treatment plants for sewage in all coastal cities by 2000;
- adoption of measures to control pollution from vessels using the 'Port State Control' approach with meaningful incentives and penalties;
- implementation of packages of measures for dumping, transboundary movement of hazardous waste, emergency response, etc.;
- publication of five-yearly 'State of the Black Sea' reports based upon a joint monitoring system and regular surveys of land-based sources of pollution;
- introduction of a regime for the joint control of fisheries based upon a quota system for capture and the rehabilitation of key ecosystems which act as nursery grounds;
- development of a new Protocol on Black Sea Biological and Landscape Diversity;
- the design of a comprehensive package of investments in conservation areas (wetlands and marine ecosystems);
- the introduction of compulsory Environmental Impact Assessments, Environmental Audits and Strategic Environmental Impact Assessments with harmonised criteria;
- development of a regional strategy for Integrated Coastal Zone Management with associated legal instruments by 1999;

- economic development in the areas of sustainable aquaculture and tourism;
- increased attention to public participation, based upon a comprehensive package including local authorities, NGOs, private sector, Regional Environmental Centres, schools, etc.;
- enhanced transparency through rights of access to information and improved public awareness;
- continuation of the BS-SAP process at the national level through the development of National Black Sea Strategic Action Plans;
- organisation of five-yearly donor meetings including the development of blended packages of investment based upon revisions of the NBS-SAPs;
- completion of a feasibility study of the Black Sea Environmental Fund which could be supported by regionally-applied economic instruments and would address incremental costs;
- regular and transparent revision of progress on implementation and the updating of the SAP objectives.

[1] Adopted at the Ministerial Conference held at the City of Istanbul, 31 October, 1996. The full text is available from the Black Sea Environmental Programme, II Harekat Kosku, Dolmabahce Sarayi, Besiktas, 80680, Istanbul, Turkey.

Profiles of contributors

Editors

Sarah Hobson

Sarah Hobson is a communication specialist whose work is dedicated to giving voice to people in disadvantaged communities. Staying in villages in Iran, India, West Africa and the Andes, she has developed an understanding of how people combine their sense of the environment with the practical needs of survival, within the framework of their religion. She has made more than ten television documentaries and published books and articles that have influenced government and donor policies. She is currently Director of Open Channels, an international non-profit making organisation that works in partnership with and in support of community and non-profit organisations representing the most disadvantaged, to improve their communication and information capability on a planned and sustainable basis.

Laurence Mee

Professor Laurence David Mee is a specialist in environmental science and policy. Born in Ipswich, England, he received a PhD in Chemical Oceanography from the University of Liverpool in 1977. His postgraduate research included work off the West Coast of Africa and in Mexico where he studied coastal lagoon ecosystems. Dr. Mee worked for 10 years as a researcher and Professor in the National Autonomous University of Mexico in Mazatlan and Mexico City, travelling extensively in Latin America to research in marine fertility and geochemistry. In 1987 he became Head of the Marine Environmental Studies Laboratory, IAEA MEL, Monaco, a unit responsible for providing technical support to the UN system for regional pollution assessments. This enabled him to study the impact of pollution in many settings including Central America, the Gulf (the impact of the Gulf war), the Caribbean, Mediterranean, Danube River and Black Sea as well as to train young specialists from many parts of the world. In 1991, he participated in the group which prepared the Chapter on Oceans and Coastal Areas of 'Agenda 21', the final document of the Rio Summit. His widening interest in social as well as ecological aspects of marine degradation took him to the Black Sea in 1991 where he assisted the governments to draft the Odessa Ministerial Declaration on the Protection of the Black Sea, and to obtain $9.3 millions from the newly-formed Global Environment Facility for a three-year Black Sea Environmental Programme (BSEP) to improve technical capacity, environmental policy and investments in the region. In 1993 he was appointed by UNDP as Coordinator of the BSEP and moved to Istanbul. The Programme enabled the drafting of the Black Sea Action Plan, a long-term strategy of profound reforms adopted by all six governments of Black Sea countries. Dr. Mee is a Fellow of the Royal Society of Chemistry, an Academician of the Ukrainian Academy of Ecological Sciences, Honorary Member of the Georgian Academy of Ecological Sciences and Chairman of the Board on Coastal and Ocean Management of the Advisory Committee on the Protection of the Sea. In 1998, the Year of the Oceans, he became visiting Professor at the University of Plymouth, UK, and was distinguished as a Pew Fellow for Marine Conservation, an award which will allow him to continue his practical activities to advance environmental education in the Black Sea region and to explore the societal causes and solutions to Black Sea degradation. He has traveled to more than 35 countries in the course of his work and has over 60 scientific publications.

Contributors

Sadruddin Aga Khan

Sadruddin Aga Khan was born in Paris in 1933. He graduated from Harvard University in 1954, with post-graduate work at the Centre for Middle Eastern Studies, Harvard, 1955-57. He was appointed UNESCO Special Consultant in 1958, and in succession served as Head of Mission and Advisor to the UN High Commissioners for Refugees, 1959-60; Special Advisor to the Director-General of UNESCO for the Organisation of the Campaign for the Preservation of Nubian Monuments; subsequently, Executive Secretary of the UNESCO International Action Committee for the Preservation of Nubia, responsible for the coordination of international archaeological excavations, 1961; UN Deputy High Commissioner for Refugees, 1962-65; and as UN High Commissioner for Refugees, 1965-77, in which capacity he was Coordinator, Special Operations for Displaced Persons in Bangladesh 1972, Uganda 1972, Sudan 1973, Chile 1973, and Cyprus 1974. He has been Special Consultant and Chargé de Mission to the UN Secretary-General since 1978. Special Rapporteur of the UN Human Rights Commission, 1981, Sadruddin Aga Khan was subsequently Convenor and Co-Chairman first of the Independent Commission on International Humanitarian Issues, 1983-87, and then of the Independent Working Group on the UN Financial Emergency, 1986; Member of the International Panel of the United Nations Association (USA): Decision-Making Project on the Future of the UN, 1987; Coordinator for UN Humanitarian and Economic Assistance Programmes Relating to the People of Afghanistan, 1988-90; Personal Representative of the UN Secretary-General for Humanitarian Assistance Relating to the Crisis between Iraq and Kuwait, 1990-91; Executive Delegate of the Secretary-General for a UN Inter-Agency Humanitarian Programme for Iraq, Kuwait and the Iraq/Turkey and Iraq/Iran border areas, 1991. Since 1992 he has been Chargé de Mission to the UN Secretary-General. He has written many articles in journals and newspapers on disarmament and environmental questions. Editor of *The Paris Review*, he has also edited *Energie Nucléaire et Société,* Pergamon, 1979; *Energie et Société,* Pergamon, 1981; *Nuclear War, Nuclear Proliferation and their*

Consequences, Oxford, 1986; *Non-Proliferation in a Disarming World,* Bellerive Foundation, 1991; *International Protection of Refugees,* The Hague Academy of International Law; as well as *Fundamental Freedoms in any part of the World, with particular reference to colonial and other dependent countries* - a study on human rights and massive exoduses for the UN Commission on Human Rights. Numerous honours and distinctions include Commandeur de la Légion d'Honneur; Commander of the Golden Ark; the United Nations Human Rights Award; and the Onassis Foundation Olympia Prize. Sadruddin Aga Khan holds Honorary Doctorates at the Fletcher School of Law and Diplomacy, Tufts University; the University of Nice and the University of Leeds.

Boris Alexandrov

Boris Georgievich Alexandrov was born in 1958 in Odessa, Ukraine. A hydrobiologist with an active interest in marine biology and ecology, he received his Master's from Odessa State University, 1980, and his PhD in Hydrobiology from the Institute of Biology of Southern Seas, Sevastopol, 1988. Director since 1994 of the Odessa branch of IBSS, he had joined the branch as an engineer-biologist, 1980-86, became Junior Research Associate, 1986-88, was Secretary of the branch's Academic Council, 1989-90, was appointed Deputy Director, 1990-94, and has since 1991 also been Chief of Laboratory of theoretical fundamentals for hydrobiological amelioration. His main fields of experience are in ecological aspects of the development and growth of marine invertebrates; the study of phyto- and zoocoenoses of sea coastal zones and determination of their role in the formation of water quality; and conservation of biodiversity in the Danube Delta and Black Sea region. Dr. Alexandrov is a Corresponding Member of the Ukrainian Ecological Academy of Sciences, and a Member of the Ukraine National Committee for the UNESCO Programme on Man and the Biosphere. He has published over 70 scientific papers and co-authored: *Results of the second joint US-USSR Bering Sea expedition, summer 1994,* ed. P.F. Roscigno, US Fish Wildl. Serv. Biol. Rep. 90 (13), 1990, pp. 149-163; *Investigations of the Baltic Sea ecosystem,* Gydrometeoizdat, Leningrad 1990, pp. 136-144; *Result of the third joint US-USSR Bering & Chukchi Seas expedition (BERPAC),* summer 1988, ed. P.A. Nagel, US Fish Wildl. Serv., 1992, pp. 184-191; *Molismology of the Black Sea,* Naukova Dumka, Kiev 1992, pp. 50-62, in Russian; and *Report on the biodiversity of the Ukrainian Black Sea Area*, comp. B.G. Alexandrov. Yu. P. Zaitsev, GEF Black Sea Environmental Programme, Odessa 1996, 290pp.

Daniel Amit

Professor Daniel Amit was born in Lodz, Poland in 1938 and received his PhD in Physics from Brandeis University in 1966. Professor of Physics at the University of Jerusalem and the University of Rome, La Sapienza. Chairman, Racah Institute of Physics, 1984-87.

Author of *Fields Theory and Critical Phenomena (*McGraw-Hill), and *Modelling Brain Function* (Cambridge).

Neal Ascherson

Neal Ascherson was born in Edinburgh, the son of a naval officer, and took a History degree at Cambridge. He has spent most of his life as a journalist, principally as a foreign correspondent or writer on foreign affairs. In the 1960s, he was *The Observer's* correspondent in Germany and east-central Europe, spending much time in Poland and Czechoslovakia. He covered the emergence of the West German Ostpolitik, the Prague spring of 1968 and the Soviet-led invasion, and the birth of Solidarity in Poland in 1980. In 1975 he returned to Scotland, and worked for five years as the Scottish politics correspondent of *The Scotsman.* He also worked outside Europe, covering events like the Namibian war, the transformation of Rhodesia and Zimbabwe, conflicts in South Africa and the Pinochet putsch in Chile which destroyed the regime of Salvador Allende. From 1985 onwards, he contributed a weekly column to *The Observer*, which he transferred to the *Independent on Sunday* when he joined that paper at its foundation in 1989. In August 1991, he attended the International Congress of Byzantinologists in Moscow, and was on a post-congress study tour of Crimea when the *coup d'état* against Mikhail Gorbachev took place. He has been awarded a number of prizes for journalism and literature, including the James Cameron Award, the George Orwell Award for Political Writing and the Journalist of the Year prize for 1987. Ascherson's first book was *The King Incorporated,* 1963, a study of Leopold II and the Congo Free State. *The Polish August* appeared in 1981, and *The Struggles for Poland* in 1987. *Games with Shadows*, a selection of his columns and shorter writing, came out in 1988. *Black Sea*, a study of the history, archaeology and ecology of the Black Sea region which concentrated on myths and mysteries of national identity, was published in 1995.

Zaki Badawi

Dr Zaki Badawi is Principal of The Muslim College, London, a post-graduate institution for the training of Islamic scholars. He is also Chairman of the Imams and Mosques Council and of the Muslim Law (Shariah) Council, both in the UK, and Vice-Chairman of the World Congress of Faiths. Educated at the Al-Azhar University, Cairo - Al Aliyah & Al Alimiyyah, and with a BA and PhD from London University, he was Director of the Islamic Banking System, Luxembourg, 1982-86, and was on its behalf a member of the negotiating committee with the Bank of England which led to permission to establish the Islamic Finance House in London in 1983, of which Dr. Badawi was appointed Manager, 1983-86. In 1996 he was Shariah Consultant to the *Islamic Banker* magazine. Earlier, as a Lecturer at Al-Azhar University, he helped plan establishment in Malaysia of the Higher Islamic Institution, later the National University of Malaysia, 1955-56. He was a Lecturer in Islamic Studies and Arabic, University of Malay in Singapore, 1956-60; moved to the University of Malaya in Kuala Lumpur, 1960-64; was Professor of Islamic Studies, and later Dean of Arts and Islamic Studies, Ahmadu Bello University, Nigeria, 1964-76; Professor of Islamic Studies, King Abdul Aziz University, Jeddah, Hajj Research Centre, 1976-78; and Director and Chief Imam of the Islamic Cultural Centre and Regent's Park Mosque, 1978-81. Dr. Badawi is also a Member of The High Council of Islamic Affairs, Egypt; The World Council for Islamic Call, Libya; the Mu'tama Al Alam al Islami, Pakistan; and the Inter-Religious Federation for World Peace. He frequently broadcasts on Muslim affairs, his areas of interest being Islamic theology and thought, Muslim jurisprudence, and the history of Muslim law.

Bishop Basil of Sergievo

Bishop Basil (Osborne) of Sergievo was born in Egypt in 1938, but grew up and was educated in the United States, where he took a doctorate in Classics at the University of Cincinnati. In 1962 he was received into the Russian Orthodox Church. He was ordained deacon in 1969 and priest in 1973, since when he has served as rector of the Russian parish in Oxford. After the death of his wife, he was consecrated titular Bishop of Sergievo in 1993 and since that time has also served as an assistant bishop in the Diocese of Sourozh (Great Britain). Bishop Basil is the editor of the quarterly journal *Sourozh*, and is the author of two small books of sermons, *The Light of Christ* and *Speaking of the Kingdom.*

Arabinda Basu

Professor Arabinda Basu is editor of *Gavesna*, a review of philosophy and yoga published by the Sri Aurobindo International Centre of Education, Pondicherry, India. He was formerly Sri Aurobindo Professor of Philosophy, Banaras Hindu University, Varanasi, UP, India, and a Lecturer in Philosophy and later in Indian Philosophy and Religion at the same university, as well as Spalding Lecturer in Indian Philosophy and Religion, University of Durham, UK, Visiting Professor at the Hebrew University of Jerusalem, the University of Dubuque, Iowa, and the Punjabi University, Punjab, India, and Visiting Lecturer at the School of Theology, Claremont Colleges, Claremont, USA. Has also worked as a programme assistant at the Calcutta station of All India Radio, and as sub-editor of *The Statesman*, Calcutta. Member of the Committee of Experts on Translation of Representative Works, International Council for Philosophic and Humanistic Studies, Paris, 1954-55. He has published three books, and articles inter alia in *Encyclopaedia Britannica; History of Philosophy Eastern and Western,* published for the Government of India by George Allen & Unwin; and *The Cultural Heritage of India,* published by the Ramakrishna Mission Institute of Culture, Calcutta.

Nicoloz Beroutchachvili

Nicoloz Beroutchachvili was born in 1947 in Tbilisi, Georgia. He graduated from Tbilisi State University, faculty of Geography-Geology, in 1970, obtaining his Candidate of Sciences degree from Moscow State University in 1971 and his doctorate in Geographical Sciences from the same university in 1981. Since 1982 he has held the chair of Cartography at Tbilisi University, where he was appointed Professor in 1983, and has since 1991 held the chair of Cartography and Geoinformation. Professor Beroutchachvili completed a special course at Paris VII University, 1975-76, and was invited back to give lectures and pursue scientific work there in 1982, 1990 and 1992-93. He has also lectured at other European universities, including those in Warsaw, Sofia, Bruno, Tuluze and Grenoble. He has 150 scientific publications to his credit, 28 of these during the past five years and including 16 monographs, two of which were published in Paris and three in Moscow. He is a member of a Russian Academy of Sciences unit concerned with fundamental geographical problems; of an FSU universities scientific methodology unit; and, since 1995, of the Georgian Academy of Natural Sciences. He is also a correspondent of the magazine *Nature, Science, Société,* Paris. Professor Beroutchachvili has organised numerous all-union and international seminars, conferences and congresses and presently chairs the cartography and scientific research laboratory at the station of Martkopi.

Patricia Birnie

Professor Patricia Birnie is currently Chair of the Legal Committee of the Advisory Committee on Protection of the Sea (ACOPS). BA (Hons) Jurisprudence (Oxon) 1947; PhD Edinburgh 1979. Barrister (Gray's Inn), 1952. Tutor / Lecturer in Public International Law, University of Edinburgh, 1967-83; Lecturer / Senior Lecturer in Law, London School of Economics, 1983-89; Professor and first Director of the International Maritime Organisation (IMO), International Maritime Law Institute, Malta, 1989-92. She has published widely, including, with R. Barston as editor and contributors, *The Maritime Dimension*, Allen & Unwin, 1980; P. Birnie, *The International Regulation Whaling*, 2 vols, Ocean, Dobbs Ferry, New York, 1985; and with A. Boyle, *International Law and the Environment,* Clarendon Press, Oxford, 1992. She has also contributed chapters to numerous books and learned journals, given papers at many international conferences, especially in her fields of special interest, the Law of the Sea and International Environmental Law, and undertaken various international consultancies.

Ritt Bjerregaard

Ritt Bjerregaard was born in Copenhagen in 1941. She has been the EU Commissioner for the Environment and Nuclear Safety since January 1995, but her political career was launched back in 1971, when she became a member of the Danish Parliament for the Social Democratic Party. After two years in Parliament she became Minister of Education, an appointment which she held until 1978, only to move to the Ministry of Social Affairs, where she was Minister from 1979 to 1981. Ms. Bjerregaard has twice been chairperson of The Parliamentary Group of the Social Democratic Party, from 1981-82 and again 1987-92. She was deputy chairperson of the same group, 1982-87, and she has been chairperson of the Parliamentary Committee on Public Accounts since 1990, as well as being during her political career a member of Parliamentary Committees on Foreign affairs, European Affairs, the Environment and Budget. At a European level, Ms. Bjerregaard has been Vice-President of the Parliamentary Assembly of the Conference on Security and Cooperation in Europe -CSCE, since 1992. She has also since 1992 been Vice-President of Socialist International Women, while continuing to participate in the political and public debate as an author of several books on politics and the role of women in politics.

John Broome

John Broome is Professor of Philosophy at the University of St. Andrews, and was previously Professor of Economics and Philosophy at the University of Bristol. He works in areas where economics and ethics come together; he applies formal techniques from economics to ethical theory, and he examines ethical questions raised by economics. One of his interests - combining ethics and economics - is the value of the environment. Among his books are *Weighing Goods,* Blackwell, 1991 and *Counting the Cost of Global Warming*, White Horse Press, 1992. He is now working on *Weighing Lives*, which deals with the value of extending people's lives and the value of adding new lives to the population. At present, he is Visiting Fellow at the Swedish Collegium for Advanced Study in Social Science, at Uppsala.

Richard Chartres

The Rt. Revd. and Rt. Hon. the Lord Bishop of London, Richard John Carew Chartres was installed as the 132nd Bishop of London in January 1996 at St. Paul's Cathedral. He was educated at Hertford Grammar School and studied History at Trinity College, Cambridge, before undertaking his theological training at Cuddesdon College, Oxford and Lincoln Theological College. Ordained deacon in 1973 and priested in 1974, he served as curate of St. Andrews, Bedford before becoming Domestic Chaplain to the Bishop of St. Albans, Robert Runcie. From 1980 to 1984 he was Chaplain to Robert Runcie, by then Archbishop of Canterbury. In 1984, Richard Chartres became Priest-in-Charge at St. Stephen's, Rochester Row (St. Stephen with St. John Westminster), and later vicar of that parish. He was Director of Ordinands for the Central London Area of the London Diocese from 1985 to 1992, and Professor of Divinity at Gresham College (a postgraduate institute founded by Sir Thomas Gresham in 1957 and run jointly by the City Corporation Mercers Company) from 1986 to 1992. He was consecrated Bishop of Stepney in 1992. Richard Chartres has a strong interest in the life and culture of London and in particular the East End. As Bishop of Stepney he was actively involved in the development programme for the Tower Hamlets Mission for Drug and Alcohol Abuse and also an active fundraiser for the Tower Hamlets Old People's Welfare Trust and many other local voluntary charities and organisations. Richard Chartres is also Ecclesiastical Patron of the Prayer Book Society and a number of other national organisations. He is a trustee of the St. Catherine's Foundation. The Foundation was set up at the instigation of HRH The Prince of Wales and brings together distinguished international religious figures, academics, businessmen and others united in their concern for the preservation of the buildings, collections and traditions of the monasteries of St. Catherine at Mount Sinai. His interests include the art and literature of the ancient Mediterranean and he has been a regular visitor to Russia over many years. He is now Chairman of the Archbishop of Canterbury's Liaison Group for the Russian Orthodox Church.

Severn Cullis-Suzuki

Severn is a 17-year-old student from Vancouver, Canada. Severn has been active in environmental work since kindergarten. She has worked with native peoples in British Columbia, south-east Asia and the Amazon to protect threatened forests from logging, and has been adopted into the Raven clan of the Haida Nation with the name Killthgula Gaayaa, Good Speaker. She started the Environmental Children's Organisation, a small group which raised funds to come to the Rio Earth Summit in 1992 'to act as a conscience to the decision-makers'. Speaking there in many venues - the Global Forum, the Earth Parliament, the UNCED panel on Ethics in Education, and a UNICEF press conference - the group achieved this goal when Severn closed a plenary session of the Earth Summit with a powerful speech. She has since given countless speeches to schools, corporations, conferences and international gatherings around the world on the necessity of changing our values, of listening to children, and of behaving as if their future matters. She also works to encourage young people to speak out. She is an accomplished television host and presenter, having participated in many programmes and series in Canada, the US, and Britain. She has written many articles on environmental issues for magazines and newspapers and has published a book. Severn received the Global 500 Award in Beijing in 1993.

Herman Daly

Dr. Herman E. Daly is currently Senior Research Scholar at the University of Maryland, School of Public Affairs. From 1988 to 1994 he was Senior Economist in the Environment

Department of the World Bank. Prior to 1988 he was Alumni Professor of Economics at Louisiana State University, where he taught economics for 20 years. He holds a BA from Rice University and a PhD from Vanderbilt University. He has served as Ford Foundation Visiting Professor at the University of Ceará, Brazil; as a Research Associate at Yale University; as a Visiting Fellow at the Australian National University; and as a Senior Fulbright Lecturer in Brazil. He has served on the boards of directors of numerous environmental organisations, including the International Society for Ecological Economics, the Beijer Ecological Economics Institute of the Swedish Royal Academy of Sciences and the WorldWatch Institute, and is co-founder and associate editor of the journal *Ecological Economics*. His interest in economic development, population, resources and environment has resulted in over 100 articles in professional journals and anthologies, as well as numerous books, including *Toward a Steady-State Economy*, 1973; *Steady-State Economics*, 1977, 1991; *Valuing the Earth,* 1993; and *Beyond Growth*, 1996. He is co-author with the theologian John B Cobb, Jr of *For the Common Good*, 1989, 1994, which received the 1991 Grawemeyer Award for Ideas for Improving World Order. In 1996 he received the Honorary Right Livelihood Award (the 'alternative Nobel prize'), and the Heineken Prize for Environmental Science awarded by the Royal Netherlands Academy of Arts and Sciences.

Sylvia Earle

Dr. Sylvia Earle is a marine biologist, explorer, author, lecturer, corporate executive and spokesman for ocean conservation. She is the Founder of Deep Ocean Engineering, Inc., Founder and Chairman of Deep Ocean Exploration and Research Inc., Senior Advisor to Sea Web, Advisor to the Independent World Commission on the Oceans, and formerly the Chief Scientist of the National Oceanic and Atmospheric Administration. Her research concerns the ecology of marine ecosystems and the development of technology for access and research in the deep sea. She serves on the board of Dresser Industries, Oryx Energy Inc., and various scientific, technical and conservation organisations. Concerned about deterioration she has observed in the ocean during more than 6,000 hours of diving and aware that most of the world's oceans remain unexplored, she has become an active advocate of exploration, research and protection for marine ecosystems. Holder of several diving records, author of more than 100 scientific, technical and popular publications, participant in numerous television programmes and leader of more than 50 expeditions, Dr. Earle has a BS from Florida State University, MA and PhD from Duke University, nine honorary doctorates and various honours including the Order of the Golden Ark by the Prince of the Netherlands, the Explorers Club Medal, Lowell Thomas Award, the Charles A. and Ann Morrow Lindbergh Award, the Bal de la Mer Sea Keeper Award, Marine Technology Society Compass Award, Julius B. Stratton Leadership Award, Boston Museum of Science Bradford Washburn Award, Gold Medal,

Society of Women Geographers, Radcliffe College Alumni Medal, Natural Resources Council Director's Award and US Department of Interior Conservation Service Award.

Charles Ehler

Charles N. ('Bud') Ehler is the Director of the Office of Ocean Resources Conservation and Assessment (ORCA) within the National Oceanic and Atmospheric Administration (NOAA), the civilian ocean agency of the United States of America. He is responsible for managing NOAA's coastal assessment and monitoring programmes, including oil and hazardous materials spill response, natural resource damage assessment, coastal pollution monitoring and strategic assessments of coastal ocean resource use and conservation. A recognised international expert on integrated coastal management, he participated actively in preparing the United Nations Environment Programme's (UNEP) Global Programme of Action for Protection of the Marine Environment from Land-Based Activities, and is a member of the Coastal Zones and Small Islands panel of the Intergovernmental Panel of Climate Change (IPCC). He guided NOAA's scientific activities during the $1 billion response to the *Exxon Valdez* oil spill in Alaska, and was deeply involved in preparing the natural resource damage assessment and negotiating the $2 billion settlement against the Exxon Corporation for damages to natural resources caused by the spill. He has worked for NOAA for almost 20 years, for the US Environmental Protection Agency's Office of Research and Development, and taught Regional Planning and Natural Resources Management at the University of Michigan, the University of California at Los Angeles, and the State University of New York at Stony Brook.

Paul Ehrlich

Professor Paul R. Ehrlich was born in Philadelphia in 1932. AB 1953, University of Pennsylvania; MA 1955, PhD 1957, University of Kansas, he is presently Bing Professor of Population Studies, Department of Biological Sciences, Stanford University, where he has been a member of the faculty since 1959. He has done field, laboratory and theoretical research on problems ranging from the dynamics and genetics of insect populations and the ecological and evolutionary interaction of plants and herbivores to experimental studies of the effects of crowding on human beings. Author of more than 700 scientific papers and articles in the popular press and 30 books, including *The Population Bomb, The Process of Evolution, Ecoscience, The Machinery of Nature, Extinction, Earth, Science of Ecology, The Birder's Handbook, New World/New Mind, The Population Explosion, Healing the Planet, Birds in Jeopardy, The Stork and the Plow,* and *Betrayal of Science and Reason.* Dr Ehrlich is the Honorary President of Zero Population Growth, and was President of the American Institute of Biological Sciences. He is a Fellow of the American Association for the Advancement of Science and the

American Academy of Arts and Sciences, and a Member of the United States National Academy of Sciences and the American Philosophical Society. Among his many honours was the first AAAS/Scientific American Prize for Science in the Service of Humanity; the Royal Swedish Academy of Sciences, Crafoord Prize in Population Biology and the Conservation of Biological Diversity (given explicitly as a substitute for the Nobel Prize in areas where the Nobel is not given); a MacArthur Prize Fellowship; the Volvo Environment Prize; The International Centre for Tropical Ecology, World Ecology Medal; International Ecology Institute Prize; UNEP Sasakawa Environment Prize; and the Heinz Award for the Environment. Dr. Ehrlich has appeared on many TV and radio programmes; he was a correspondent for NBC TV, and has given hundreds of public lectures in the past 25 years.

Oktay Ekinci

Born in Balikesir in 1952, Oktay Ekinci is a graduate of the Architecture Department of the State Fine Arts Academy, Turkey. He was Deputy President of the Turkish Chamber of Architects between 1988 and 1990 and Chairman of the Istanbul Branch of the Turkish Chamber of Architects between 1992 and 1994. He teaches on city conservation at the Mimar Sinan University. He has been elected President of the Turkish Chamber of Architects and writes articles for Cumhuriyet newspaper.

Edward Goldsmith

Edward Goldsmith was born in Paris in 1928, holds dual British and French nationality and was educated at Millfield School and Magdalen College. Oxford. He has an MA in Politics, Philosophy and Economics. He has been Publisher of *The Ecologist* since 1969, and was also its Editor from 1969 to 1990. He is author of *The Stable Society*, 1978; *The Great U-Turn*, 1988; and *The Way: An Ecological World-View*, 1992; co-author of *A Blueprint for Survival*, 1972, Vol I of *The Social and Environmental Effects of Large Dams*, 1984, and *5,000 Days to Save the Planet*, 1990; editor of *Can Britain Survive?*, 1971; and co-editor of Vols II and III of *The Social and Environmental Effects of Large Dams*, 1986, 1992, *The Earth Report*, 1988, *Gaia, the Thesis, the Mechanisms and the Implications*, 1988, and *Gaia and Evolution*, 1990. Edward Goldsmith taught courses on Ecology and related subjects at Michigan University, 1975, and Sangamon State University, 1984, and is still working with a Global Ecology Course which he set up in 1990 with the International Honours Programme, associated with Bard College. He is Vice-President of Ecoropa, and President of Ecoropa, France, a Member of the Praesidium of the Vienna Academy for Future Studies, Director of the Schumacher Society, Principal Environmental Consultant to the Goldsmith Foundation, and Trustee of the Foundation for Gaia. He received the Honorary Right Livelihood Award (known as the alternative Nobel Prize) in 1991 and was in the same year made a Chevalier de la Légion d'Honneur.

Sylvie Goyet

Sylvie Goyet's professional experience originates from business management and economics. Further to completing a Master's degree in International Business from the Graduate Business School of Lyon, France, and an MBA in Finance from the University of Texas, she worked as a Financial Comptroller and Commercial Manager for private sector companies in Europe. Moving into development work and natural resources management, she then obtained a Master's degree in Environmental Management from the University of London and joined the United Nations Development Programme, Fiji regional office. Programme Officer for North Pacific Island Countries, she managed the UNDP country programme for the Marshall Islands, FSM and Palau and led the development and implementation of projects and activities, in particular in the area of natural resources planning and sustainable development. She joined the GEF - Black Sea Environmental Programme (BSEP) in March 1993, as Institutional Development Officer. She was there responsible for coordinating the Integrated Coastal Zone Management programme, for initiating activities with regard to the development of sustainable tourism practices in the Black Sea coastal areas, and for stimulating the involvement and participation of NGOs in the BSEP. In 1997 she took up the position of Programme Officer at WWF International in Gland, Switzerland. She oversees WWF programmes in Indonesia and the South Pacific and leads the monitoring/evaluation and strategic planning efforts at WWF Asia/Pacific regional bureau. Through her academic, professional work and consultancy assignments, she has become experienced in coastal zone management programmes and policies, environmental public awareness activities and NGO strengthening as well as overall environmental project management. Thanks to her work and travels, she has gained familiarity with Western and Eastern Europe, North America, South East Asia and the South Pacific, Mongolia and Africa.

Richard Harbison

Dr. Richard Harbison is a biologist who studies the ecology and natural history of planktonic animals. He received his BA from Columbia University in 1966, and his PhD from Florida State University in 1971. He has been a Principal Research Scientist at the Australian Institute of Marine Science, Director of Marine Science at the Harbor Branch Oceanographic Institution, and is presently a Senior Scientist at the Woods Hole Oceanographic Institution. He has participated in numerous field expeditions ranging from the Arctic and the Antarctic to the tropics. Recently, he has been concerned with the ways in which individual species can drastically alter ecosystems and thereby affect human activities, and has been working on a solution to the problem created when an exotic planktonic animal, the ctenophore *Mnemiopsis leidyi,* was accidentally introduced into the Black Sea with devastating environmental and economic consequences.

Graeme Kelleher

Dr. Graeme Kelleher trained as a civil engineer, a profession which he practised in the design and construction of major projects in Australia and Canada for 20 years. For the past 20 years he has been engaged in environmental and natural resource management both nationally in Australia and internationally. He was Chairman and Chief Executive of the Great Barrier Marine Park Authority between 1979 and 1994, and has been Vice Chairman (Marine) of IUCN's World Commission in Protected Areas since 1986. In the latter position he has been responsible for establishing the foundation for a Global Representative System of Marine Protected Areas. This system is designed to protect habitats which represent all the major bio-geographic zones of the world's coastal seas. Among various other positions, Graeme Kelleher has occupied the Kevin Stark Chair of Systems Engineering at James Cook University, Townsville between 1992 and 1995. He is a Fellow of the Order of Australia, of the Australian Academy of Technological Sciences and Engineering and of the Institution of Engineers, Australia.

Terry De Lacy

Professor Terry De Lacy has since July 1997 been CEO of Australia's Cooperative Research Centre for Sustainable Tourism. The CRC, which is a partnership between industry, universities and government, has a vision of a dynamic, internationally competitive and environmentally sensitive tourism industry, and a mission of delivering strategic knowledge and products to business, community and government to enhance the environmental, economic and social sustainability of tourism. Previously Terry De Lacy was foundation Dean of the Faculty of Land and Food Systems and Professor of Environmental Policy, University of Queensland. He has co-authored or edited four recent texts on indigenous land management, non-market valuation of forests, integrated conservation and development, and sustainable management of grass lands in Inner Mongolia, China. He has been commissioned by Environment Australia to write a text on protected area management, and by UNESCO to write a text in the MAB series on his integrated research on conservation and development in Asia. He has received three Australian Research Council large grants since 1992. His research area is in environmental policy analysis, specialising in ecologically sustainable development. He is currently leading a team undertaking a comparative study in Asia and the Pacific on linking development and conservation through biosphere reserves, and has recently completed an evaluation of landscape groups in Australia for the Australian Department of Primary Industries and Energy. He has extensive research experience in China (Inner Mongolia, Fujian and Guandong) working with the Chinese Academy of Science and Ministry of Forestry; in Nepal (Makalu and Terai area) working with the Nepal Department of National Parks and Wildlife Conservation; and more recently, in Indonesia (East Java) working with Gadja Marta University and in Thailand working with colleagues at Kasetstart

University. He is a member of the Board of Australian Cotton Research and Development Corporation, a Director of the South East Asian Consortium on Research and Postgraduate Education in Tropical Agriculture and Natural Resources, and an Australian delegate to IUCN's International Commission on Parks and Protected Areas.

Jane Lubchenco

Wayne and Gladys Valley Professor of Marine Biology and Distinguished Professor of Zoology at Oregon State University, a Pew Scholar in Conservation and the Environment, and a MacArthur Fellow, Dr. Jane Lubchenco is Past-President of the American Association for the Advancement of Science, a past President of the Ecological Society of America, and a Member of the National Academy of Sciences. A marine ecologist by training, with a PhD from Harvard University, Dr. Lubchenco's current research interests include marine conservation biology, biological diversity, ecosystem services, ecological causes and consequences of global changes, and sustainable ecological systems. She is engaged in a wide variety of activities intended to help address numerous serious environmental problems by improving the scientific understanding of issues, by making the best possible scientific information and expertise more accessible to policy and decision makers, and by improving the public's understanding of ecological topics. She led the innovative efforts of the Ecological Society of America to set national priorities for ecological research, an endeavour which resulted in the Sustainable Biosphere Initiative, and she co-chairs the international SCOPE Sustainable Biosphere Project, an interdisciplinary scientific programme designed to propose environmentally sound policy and management options for seven regions of the world. Dr. Lubchenco coordinated the sections of the United Nations Environment Programme's newly released Global Biodiversity Assessment, which deals with biological diversity and ecosystem functioning. She was named Outstanding Teacher of the Year at Oregon State University in 1986, and Oregon Scientist of the Year in 1994. She is a member of the boards of directors of the World Resources Institute, The Environmental Defense Fund, and the Monterey Bay Aquarium. She serves on several advisory committees for the National Research Council, the National Science Foundation, National Public Radio's *Living on Earth*, the United Nations Environment Programme, and UNESCO.

John Lyras

John C. Lyras is a Director of Paralos Maritime Corporation SA, and has been involved in shipping since 1975 in London and 1979 in Piraeus. John Lyras comes from a Greek seafaring family with a tradition in shipping that goes back four generations. He holds a Bachelor's Degree in Social Science and a Master's degree in Transportation Planning and Management. He is the President of the Union of Greek Shipowners, having been a Member of the Board since 1981 and Vice-President since 1991. He also serves as

Chairman of the Union's Foreign Affairs Committee. Mr. Lyras is the immediate past President of the European Community Shipowners' Associations, having previously been Vice-President from June 1991. He is also a founder Member and a Board Member of HELMEPA, the Hellenic Marine Environment Protection Association. A former Vice-President of the International Chamber of Shipping, Mr. Lyras served on its Executive Committee for eight years, and is currently a Member of the American Bureau of Shipping's Greek Technical Committee and a Member of Lloyd's Register of Shipping's General Committee.

Laurence Mee

Professor Laurence David Mee is a specialist in environmental science and policy. Born in Ipswich, England, he received a PhD in Chemical Oceanography from the University of Liverpool in 1977. His postgraduate research included work off the West Coast of Africa and in Mexico where he studied coastal lagoon ecosystems. Dr. Mee worked for 10 years as a researcher and Professor in the National Autonomous University of Mexico in Mazatlan and Mexico City, travelling extensively in Latin America to research in marine fertility and geochemistry. In 1987 he became Head of the Marine Environmental Studies Laboratory, IAEA MEL, Monaco, a unit responsible for providing technical support to the UN system for regional pollution assessments. This enabled him to study the impact of pollution in many settings including Central America, the Gulf (the impact of the Gulf war), the Caribbean, Mediterranean, Danube River and Black Sea as well as to train young specialists from many parts of the world. In 1991, he participated in the group which prepared the Chapter on Oceans and Coastal Areas of 'Agenda 21', the final document of the Rio Summit. His widening interest in social as well as ecological aspects of marine degradation took him to the Black Sea in 1991 where he assisted the governments to draft the Odessa Ministerial Declaration on the Protection of the Black Sea, and to obtain $9.3 millions from the newly-formed Global Environment Facility for a three-year Black Sea Environmental Programme (BSEP) to improve technical capacity, environmental policy and investments in the region. In 1993 he was appointed by UNDP as Coordinator of the BSEP and moved to Istanbul. The Programme enabled the drafting of the Black Sea Action Plan, a long-term strategy of profound reforms adopted by all six governments of Black Sea countries. Dr. Mee is a Fellow of the Royal Society of Chemistry, an Academician of the Ukrainian Academy of Ecological Sciences, Honorary Member of the Georgian Academy of Ecological Sciences and Chairman of the Board on Coastal and Ocean Management of the Advisory Committee on the Protection of the Sea. In 1998, the Year of the Oceans, he became visiting Professor at the University of Plymouth, UK, and was distinguished as a Pew Fellow for Marine Conservation, an award which will allow him to continue his practical activities to advance environmental education in the Black Sea region and to explore the societal causes and solutions to Black Sea degradation. He has travelled to more than 35 countries in the course of his work and has over 60 scientific publications.

Radu Mihnea

Radu Mihnea was born in Tecuci, Romania in 1942 and received his BA in Biology at the University of Jassy. He is presently Coordinator of the Black Sea Environmental Programme based in Istanbul, having served from October 1997 to March 1998 as Deputy Coordinator. Prior to this he was Senior Scientist, Romanian Marine Research Institute; National Coordinator, Marine Pollution Monitoring Programme, Technical expert for the Convention on the Protection of the Black Sea against Pollution; and National Focal Point for Routine Pollution Monitoring in the GEF Black Sea Programme. His key qualifications are in marine zoobentos - hard bottom populations; toxicology - bio-assays on biological effects of acute exposure to harmful substances; and marine pollution - monitoring of marine coastal waters. His professional experience includes a six-month fellowship in the UK on a UNESCO grant in toxicology testing; projection, testing and implementation of the current monitoring programme of Romanian marine waters; toxicity testing of effluents, oil dispersants, etc; Head of Marine Protection Department (Laboratory), 1981-88; National Coordinator of Monitoring of the Black Sea Environment Subprogramme, part of the Global Environment Monitoring Programme, a common project of Eastern countries in the frame of Common Mutual Aid, 1980-88; Pollution technical expert, Romanian delegation for the Convention on the Protection of the Black Sea against Pollution; he was involved in the training course in Environmental Impact Assessment and Environmental Audit as part of the European Community PHARE programme, with field studies in the UK; coordinator of many marine pollution monitoring projects, environmental impact assessments and environmental audits on chemical plants, sewage treatment stations, offshore oil drilling and touristic resorts; documentary trips in marine pollution monitoring, USA, France, UK. He has published 22 scientific papers and is co-author of *State of Marine Environment in the Black Sea Region* - the Romanian contribution to UNEP Regional Seas Reports and Studies 124/1990.

Mauro Morelli

Bishop Mauro Morelli was born in Avanhandava, Brazil in 1935, and was ordained a priest, Saint Mary's Seminary and University, Baltimore, in 1965. He served as a parish priest, Paróquia São João Batista, Rio Claro, Brazil, 1966-74; at the National Conference of Brazilian Bishops (CNBB) as Regional Secretary São Paulo State, 1971-75; elected a bishop, auxiliary to Archdiocese of São Paulo, 1974; member of the CNBB - Regional Representative, Commission, 1975-81, and its president 1979-81; member of the CNBB - National Representative Commission 1975-81; first bishop for the Roman Catholic Dioceses of Duque de Caxias, Rio de Janeiro State, 1981 to the present. Brazilian delegate to the IV Latin-American Bishops' Conference, Santo Domingo, 1992. National Food Security Council, Presidency of the Republic, President, 1993 and 1994. Independent auditor for the National Human Rights Plan, 1996 to the present. Within and outside Brazil over 20 years of participation in workshops, conferences, speaking tours and similar, related to environment, development, hunger, human rights, constitutional law, foreign debt, agrarian reform and other social issues.

Nicos Mouzelis

Nicos P. Mouzelis was born in Athens in 1939 and lived part of his childhood in the Greek city of Lamia. He studied Economics and Sociology at the Universities of Geneva and London. He is at present Professor of Sociology at the London School of Economics, University of London. He is one of the founders of ELEMEP, a forum in London for the promotion of research on Modern Greek Society, and of PAREMVASI, a citizens' movement aiming at the development of civic rights and of a strong civil society in Greece. His major publications include *Organisation and Bureaucracy: An analysis of Modern Theories*, Routledge, 1967; *Modern Greece: Facets of Underdevelopment*, Macmillan, 1978; *Politics in the Semi-Periphery: Early Parliamentarism of late Industrialisation in the Balkans and Latin America*, Macmillan, 1988; *Post Marxist Alternatives*, Macmillan, 1990; *Back to Sociological Theory*, Macmillan, 1992; and *Sociological Theory: What Went Wrong?*, Routledge, 1995. He is a regular contributor to the Greek Sunday newspaper *To Vima*.

Ulrich Niermann

Dr. Ulrich Niermann was born in Iserlohn, Germany, in 1954 and studied at the University of Hamburg, 1975-82. His basic subjects were biology, chemistry, physics and he specialised in Marine Biology and Ecology. For his PhD - thesis entitled 'Fluctuation of Macrofauna in the South-eastern North Sea 1984-1988' - he investigated the impact of eutrophication and oxygen deficiencies on the bottom fauna. From 1982 until 1990 Dr. Niermann worked as a marine biologist on various international projects at the Biologische Anstalt Helgoland in Hamburg. His main research fields were 'Reproductive biology of the eel in the North Atlantic' and long-term fluctuation of marine communities. From 1991 to 1996 Dr. Niermann was a member of the teaching staff at the Middle East Technical University, Institute of Marine Sciences in Erdemli, Turkey, delegated as an integrated expert by CIM/GTZ, Germany. He lectured on marine ecology and worked on the ecology of the zooplankton and ichthyoplankton of the Black Sea. He was involved in the international TU-Black Sea Project as one of the coordinators for the biological investigations. Since 1995 he has been a member of the Ukrainian-Turkish working group on 'long-term fluctuation of zooplankton in the Black Sea', a project which he initiated. Before coming to Turkey, Dr. Niermann was a member of the Ecology Benthos

Working Group of the ICES (International Council for the Exploration of the Sea), and one of the coordinators of the ICES International North Sea Survey, 1986-88, which was carried out with seven research vessels from different European countries for the first time in the history of North Sea research activity. Dr. Niermann has taken part in more than 40 scientific expeditions - on some 15 cruises as scientist in charge - headed to the Arctic, Antarctic, North Sea, Atlantic, Mediterranean and Black Sea, and he has published about 60 scientific publications in scientific and public journals.

Erdal Özhan

Professor Erdal Özhan is a graduate of the Middle East Technical University, Ankara, BSCE, 1970, and the Asian Institute of Technology, Bangkok, M.Eng, 1972; D. Eng, 1975. He started his academic career at the Middle East Technical University in 1975. During the following 21 years at METU, he contributed to the creation of a well-resourced graduate programme in the field of coastal engineering; led establishment of the Coastal and Harbour Engineering Research Centre (KLARE), 1978; set up the METU Subaqua Society, 1985, to involve students in protection of nature and environment in the coastal zone; and initiated and offered coastal zone management courses, 1989. In 1990, Professor Özhan started the process of establishing the Turkish National Committee on Coastal Zone Management, which was set up as an official entity in January 1993. That same year, he launched the regional MEDCOAST initiative on the Mediterranean and the Black Sea, to contribute to environmental management and sustainable development of the coastal and marine areas of these two major enclosed basins. MEDCOAST has received a high level of international recognition, and has started playing a measurable role in enhancing ICZM practices in the Mediterranean and the Black Sea. Professor Özhan has carried out research into various issues of engineering, management and conservation in the coastal zone, and has published extensively in all these three areas in the international media. He is an active member of numerous national and international professional and conservation societies, and serves on the editorial boards of two leading international scientific journals. Professor Özhan presently works as a full time Professor of Coastal Engineering and Management at METU. He is the Director of the Coastal Engineering Laboratory, the Chairman of MEDCOAST, and the Chairman of the Turkish National Committee on Coastal Zone Management. Professor Özhan has recently been selected as a 1997 Pew Fellow in Conservation and the Environment for his pioneering regional role in starting and leading the MEDCOAST initiative.

Metropolitan John of Pergamon

The Most Revd. Metropolitan John of Pergamon was born in 1931. He studied at the Theological Schools of the Universities of Thessaloniki and of Athens, graduating from the latter in 1955. After extensive studies abroad, he submitted his doctoral thesis to the

Theological School of the University of Athens and became a Doctor of Theology in 1965. He was for three years Secretary of the Faith and Order section of the World Council of Churches in Geneva. He taught Dogmatic Theology for three years at the University of Edinburgh, was Professor of Systematic Theology at the University of Glasgow for 14 years, and Visiting Professor, University of Geneva, London University and Grigorian University, Rome. A participant in many international scientific conferences, he has for many years represented the Ecumenical Patriarchate on international church bodies and on special ecclesiastical missions. Member of the International Committee for formal Theological Dialogue with the Roman Catholic Church, and Orthodox President of the International Committee for Theological Dialogue with the Anglican Church. Member for eight years of the Central Committee of the World Council of Churches and of the Faith and Order Committee of that body as representative of the Ecumenical Patriarch. Author of many scientific studies in various languages. Honorary Doctor of the Catholic Institute of Paris and the Orthodox Theological School, Belgrade. He was elected Member of the International Academy of Religious Science in Brussels in 1973. Appointed Metropolitan in 1986. Elected regular Member of the Athens Academy in 1993.

Ravi Ravindra

Professor Ravi Ravindra was born in India in 1939, and moved to Canada in 1966. He has a BSc and a Master of Technology from the Indian Institute of Technology, Kharagpur, a Master of Science and a PhD in Physics from the University of Toronto, and a Master of Arts in Philosophy from Dalhousie University. He held postdoctoral fellowships in Physics, University of Toronto, History and Philosophy of Science, Princeton University, and Religion, Columbia University. He was a Member of the Institute for Advanced Study in Princeton in 1977 in the School of Natural Sciences, and a Member of the Indian Institute for Advanced Study at Simla in 1978. He was the Founding Director of the Threshold Award for integrative knowledge, and chair of its international and interdisciplinary selection committees in 1979 and 1980. Recipient of several awards and research grants, he is the author of more than 100 papers in physics, philosophy and religion, and of the following books: *Theory of Seismic Head Waves*, with V. Cerveny, translated into Russian; *Whispers from the Other Shore*, translated into Portuguese; *Science and Spirit*, editor and principal author; *Krishnamurti: Two Birds on One Tree; The Yoga of the Christ,* translated into French, Portuguese, Greek and German; and *Yoga and the Teaching of Krishna*. He is a member of the Advisory Committee for the Programme of Dialogue between Science and Religion initiated by the American Association for the Advancement of Science. At present Ravi Ravindra is the Professor and Chair of Comparative Religion, Professor of International Development Studies and Adjunct Professor of Physics of Dalhousie University, Halifax, Canada.

William Reilly

William Kane Reilly is a principal with the Texas Pacific Group, an investment partnership based in Fort Worth and San Francisco, which invests in environmental, airline, banking and other companies in the United States, Latin America and Asia. He is the founder and CEO of Aqua International Partners, a $3000 million equity fund which finances water supply and wastewater treatment in developing countries. He is also a Visiting Professor at Stanford University's Institute for International Studies. From 1989 to 1993, he served as the seventh Administrator of the US Environmental Protection Agency. With 18,000 employees and a $7 billion budget, EPA is an independent federal regulatory agency charged with improving and protecting public health and the environment. While at EPA. Mr. Reilly consistently championed integration of the nation's environmental and economic agendas and sought to strengthen the role of science at EPA. He put priority on elevating attention to the protection and restoration of natural systems, preventing pollution before it is generated, enforcing environmental laws, and fostering such innovative clean-up technologies as bioremediation. He led the Agency in assessing and advancing the concerns of the emerging environmental justice movement. He played a pivotal role in crafting and securing passage of a new Clean Air bill, enacted by Congress and signed into law by President Bush in November 1990. He was President of World Wildlife Fund - US, 1985-89, and President of the Conservation Foundation, 1973-89. He was Executive Director of the Rockefeller Task Force on Land Use and Urban Growth from 1972 to 1973. From 1970 to 1972, he served as a senior staff member of the President's Council on Environmental Quality and, from 1968 to 1970, as Associate Director of the Urban Policy Centre and the National Urban Coalition. He serves on the boards of Allied Waste, American Farmland Trust, DuPont, Catalytica Combustion Systems, Clean Sites, Evergreen Holdings, The German Marshall Fund, National Geographic Society, Partners for Livable Communities, World Wildlife Fund, and the Yale University Corporation. An alumnus of Yale University, he holds a Law degree from Harvard University and a Master's degree in Urban Planning from Columbia University.

Naftali Rothenberg

Rabbi Naftali Rothenberg, PhD, is a Senior Fellow at the Van Leer Jerusalem Institute, where he co-chairs the Framework on Contemporary Jewish Thought and Identity. Rabbi Rothenberg is a graduate of the Hebrew University of Jerusalem, the IDF College for Staff and Command, The Kamenits Yeshiva-Theological Seminary of Jerusalem and the Negev Yeshiva in southern Israel. He holds several Rabbinical Ordinations degrees including Israeli Chief Rabbinate Ordination and BA, MA, and PhD in Philosophy. He has served as the Chief Rabbi of Peru, 1974-77; Rabbi of IDF School of Cadets and other IDF special divisions, 1977-85; Dean of IDF Educational Institute for Commanders,

1985-89; National Director of WZO-Torah Education for North America, NY office, 1989-93. He has published numerous articles on the subject of education, Halakha (Jewish law) and Jewish philosophy and has conducted several extensive pieces of research and studies on political science and public affairs from the Jewish sources. Rabbi Naftali Rothenberg and his family make their home in the Jerusalem suburb of Har Adar, where he serves as the town's rabbi and spiritual leader.

David Suzuki

Dr. David T. Suzuki was born in Vancouver, British Columbia in 1936. After the Second World War he grew up in London, Ontario. He graduated from Amherst College in 1958 with an honours BA in Biology and a PhD in Zoology from the University of Chicago in 1961. He held a research associateship in the Biology Division at Oak Ridge National Lab, 1961-62, was Assistant Professor in Genetics at the University of Alberta, 1962-63, and Assistant-Full Professor in Zoology at the University of British Columbia, 1963-present. He is currently in the SDRI (Sustainable Development Research Institute) at the University of British Columbia. He is a member of the Royal Society of Canada, Order of Canada, author of 28 books including *An Introduction to Genetic Analysis*, the most widely used genetics textbook in the US, and which has been translated into Italian, Spanish, Greek, Indonesian, Arabic, French and German; *Genethics: The Ethics of Engineering Life;* and 14 books for children. He hosted the TV series *Suzuki on Science, Science Magazine, The Nature of Things* and *Planet for the Taking*. On radio he originated and hosted *Quirks & Quarks* and *It's A Matter of Survival*. In 1981 he was a founding member of Scientists for Social Responsibility. He co-founded the David Suzuki Foundation, a charitable organisation that seeks concrete solutions to sustainability of local communities. He has worked actively with aboriginal communities for more than two decades and been honoured with four names and adoption by the Eagle Clan of the Haida Nation. He has received 12 honorary degrees from universities in Canada, the United States and Australia.

Yuvenaly Zaitsev

Professor Yuvenaly P. Zaitsev, born in 1924, is a marine biologist and ecologist. Chief Scientist, Odessa Branch, Institute of Biology of Southern Seas, National Academy of Sciences of Ukraine; Doctor in Biological Sciences, 1964; corresponding member of the National Academy of Sciences of Ukraine, 1969; full member of the Ukrainian Ecological Academy of Sciences, 1993; Vice-President of the Ukrainian Hydroecological Society, 1994. Professor Zaitsev has published more than 250 research papers and has presented more than 30 invited papers at international meetings. His experience related to the Black Sea ranges from biological diversity, pelagic and bottom communities in the sea and its wetlands and man-made changes in the Black Sea ecosystem to the protection of marine living resources. He has written six books, including: *Ichthyoplankton of the Odessa Gulf and adjacent Black Sea areas*, Kiev: Acad. Sc. Publs, 1959-96pp, in Ukrainian; *Marine Neustonology*, US Dept. of Commerce, Nat. Tech. Inform. Serv., Springfield Va., 1971, 207pp, translation from Russian; *This Wonderful Sea*, Odessa, Mayak Publs, 1978, 160pp, in Russian; *Dunayskye Plavni Reserve*, Odessa, Mayak Publs, 1989, 144pp, in Russian. He has also written chapters on different problems of marine biology and ecology in 14 books, published in the Ukraine, Russia, the Netherlands, UK and Turkey.

The symposium was made possible by

The Greek Ministry of Culture

Cultural Capital of Europe - Thessaloniki 1997

An Anonymous Greek Donor

Hellenic Telecomunications Organisation SA

Titan Cement Company SA

Sponsorship was provided by

European Commission DG XI
Environment, Nuclear Safety and Civil Protection

The David and Lucille Packard Foundation

Entraco Inc (Elinoil)

Commercial Bank of Greece

Support was provided by

Stavros Andreadis - Sani SA

Banca Bucuresti and the
Alpha Credit Bank Group

K. & N. Efthymiadis Industrial Co Ltd

Mr Stamos Fafalios

A. Hatzopoulos

Helexpo - Salonika International Fair

The A.G. Leventis Foundation

Ebie A. Michailidis SA

Kapniki A. Michailidis SA

National Bank of Greece

Niki SA MARINOPOULOS

Paralos Maritime Co SA

Silver & Baryte Ores Mining Co SA

Additional assistance was received from

Aegean Aviation

Ioannis G. Akkas Industrial Co Ltd

An Anonymous Turkish Donor

Delta SA

Doukas Tours

Elleniki Etairia Emfialoseos SA

Families Kedros

Mr L.T. Los

Dr & Mrs A.J. O'Reilly

Alexander S. Onassis Foundation

The symposium gratefully acknowledges

Assistance given by the Government of Greece
in waiving port fees for the Symposium vessel
in Piraeus and Thessaloniki

Assistance given by H.E. the President of Turkey
Mr Suleyman Demirel in the waiving of
port fees in Trabzon and Istanbul
and dues for the Straits

Assistance given by the Government of the
Autonomous Republic of Adjaria in waiving
port fees in Batumi and providing tour buses and guides

Assistance given by the Government of the
Russian Federation in waiving
port fees in Novorossiysk

Assistance given by the Government of Ukraine
in waiving port fees in Yalta and Odessa

Assistance given by the Government of Romania
in waiving port fees in Constanta

Assistance given by the Government of Bulgaria
in waiving port fees in Varna